INTEGER OPTIMIZATION AND ITS COMPUTATION IN EMERGENCY MANAGEMENT

Emerging Methodologies and
Applications in Modelling,
Identification and Control

INTEGER
OPTIMIZATION
AND ITS
COMPUTATION IN
EMERGENCY
MANAGEMENT

ZHENGTIAN WU
Suzhou University of Science and Technology
Suzhou, China

Series Editor

QUAN MIN ZHU

ACADEMIC PRESS
An imprint of Elsevier

Academic Press is an imprint of Elsevier
125 London Wall, London EC2Y 5AS, United Kingdom
525 B Street, Suite 1650, San Diego, CA 92101, United States
50 Hampshire Street, 5th Floor, Cambridge, MA 02139, United States
The Boulevard, Langford Lane, Kidlington, Oxford OX5 1GB, United Kingdom

ISBN: 978-0-323-95203-3

For information on all Academic Press publications
visit our website at https://www.elsevier.com/books-and-journals

Publisher: Mara E. Conner
Acquisitions Editor: Sophie Harrison
Editorial Project Manager: Sara Greco
Production Project Manager:
Erragounta Saibabu Rao
Cover Designer: Matthew Limbert

Typeset by VTeX

Working together
to grow libraries in
developing countries

www.elsevier.com • www.bookaid.org

Contents

Biography

Dr Zhengtian Wu (1986–)

Zhengtian Wu received two PhD degrees in Operations Research from University of Science and Technology of China and City University of Hong Kong in 2014. Now he is a IEEE Senior Member.

He is currently an associate professor in Suzhou University of Science and Technology, Suzhou, China. From September 2018 to September 2019 he is a visiting scholar in Department of Mechanical Engineering, Politecnico di Milano, Milan, Italy. His research interests include neural computation, neural networks, mixed-integer programming, approximation algorithms and distributed computation in emergency management.

CHAPTER 1

Distributed implementation of the fixed-point method for integer optimization in emergency management

1.1 Dang and Ye's fixed-point iterative method

In Dang and Ye's fixed–point iterative method [1,2], let

$$P = \{x \in R^n | Ax + Gw \le b \text{ for some } w \in R^p\},$$

where $A \in R^{m \times n}$ is an $m \times n$ integer matrix with $n \ge 2$, $G \in R^{m \times p}$ is an $m \times p$ matrix, and b is a vector in R^m.

Let $x^{\max} = (x_1^{\max}, x_2^{\max}, \ldots, x_n^{\max})^T$ with $x_j^{\max} = \max_{x \in P} x_j$, $j = 1, 2, \ldots, n$, and $x^{\min} = (x_1^{\min}, x_2^{\min}, \ldots, x_n^{\min})^T$ with $x_j^{\min} = \min_{x \in P} x_j$, $j = 1, 2, \ldots, n$. Let $D(P) = \{x \in Z^n | x^l \le x \le x^u\}$, where $x^l = \lfloor x^{\min} \rfloor$ and $x^u = \lfloor x^{\max} \rfloor$. For $z \in R^n$ and $k \in N_0$, let

$$P(z, k) = \{x \in P | x_i = z_i, 1 \le i \le k, \text{ and } x_i \le z_i, k+1 \le i \le n\}.$$

Given an integer point $y \in D(P)$ with $y_1 > x_1^l$, Dang and Ye [1,2] developed a iterative method presented in Fig. 1.1. It determines whether there is an integer point $x^* \in P$ such that $x^* \le_l y$.

To illustrate the method, we give an example. Consider a polytope $P = \{x \in R^3 | Ax \le b\}$ with

$$A = \begin{Bmatrix} -1 & 0 & 2 \\ 0 & -2 & 1 \\ -1 & 0 & -2 \\ 1 & 1 & 0 \end{Bmatrix}$$

and

$$b = \begin{Bmatrix} 0 \\ 1 \\ 1 \\ 0 \end{Bmatrix}.$$

Integer Optimization and Its Computation in Emergency Management
https://doi.org/10.1016/B978-0-32-395203-3.00006-X

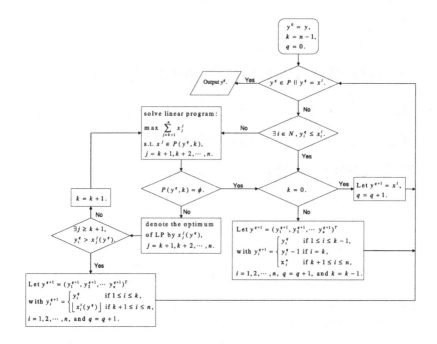

Figure 1.1 Flow diagram of the iterative method.

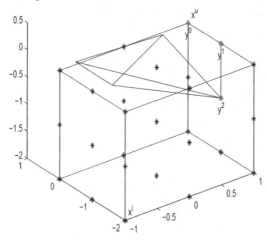

Figure 1.2 An illustration of the iterative method.

We easily obtain $x^u = (1, 0, 0)^T$ and $x^l = (-1, -2, -2)^T$. Let $y = x^u$, $y^0 = y$, and $k = 3 - 1 = 2$. In the first iteration, we obtain $y^1 = (1, -1, 0)$, and in the second iteration, $y^2 = (1, -1, -1)$, which is an integer point in P. An illustration of y^0, y^1, and y^2 is presented in Fig. 1.2.

The idea of Dang and Ye's method [1–3] to solve an integer programming problem is defining an increasing mapping from a finite lattice into itself. All the integer points outside the P are mapped into the first point in P that is smaller than them in the lexicographical order of x^l. All the integer points inside the polytope are fixed points under this increasing mapping. Given an initial integer point, the method either yields an integer point in the polytope or proves that no such point exists within a finite number of iterations. For more details and proofs about this iterative method, we refer to Dang [1,2].

1.2 Some details of the distributed implementation

As an appealing feature, Dang and Ye's method can be easily implemented in a distributed way. Some distributed implementation techniques for Dang and Ye's method will be discussed in this section.

In distributed computation, a problem is divided into many subproblems, each of which can be solved by different computers, which communicate with each other by messages. Distributed models can be classified into simple and interactive models, illustrated in Figs. 1.3 and 1.4, respectively.

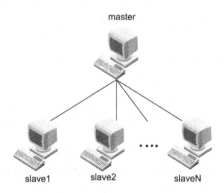

Figure 1.3 The simple distributed model.

The simple distributed model, as described in Fig. 1.3, has been used in our implementation. There are one master computer and a certain number of slave computers in this distributed computing system. The master computer takes charge of computing the solution space of the polytope, dividing the solution space to segments, sending the segments to the slave computers, receiving the computation result from the slave computers, and

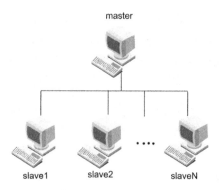

Figure 1.4 The interactive distributed model.

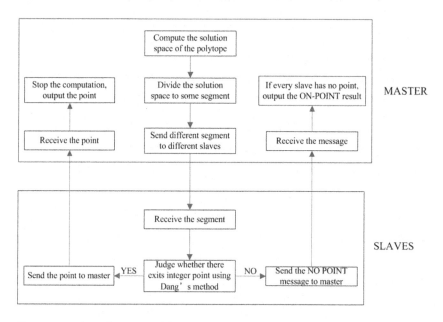

Figure 1.5 Flow diagram of the distributed computation process.

exporting the computation result. Each slave computer receives the segment, judges whether there exists an integer point in its segment using Dang and Ye's fixed-point iterative method, and sends its result to the master. The outline of the distributed computation process in this chapter is explained in Fig. 1.5.

All the programs are coded in C++ and run on Microsoft Windows platform. Two algorithms have been considered for the bounding linear program in Dang's iterative method. One is the simplex algorithm [4], the most popular algorithm for linear programming. We can call the API function in CPLEX Concert Technology to carry out this method. The other algorithm is the self-dual embedding technique presented in [5,6]. This algorithm detects LP infeasibility based on a proved criterion, and to our knowledge, it is the best method to solve linear programming problems by Dang and Ye's algorithm.

The MPICH2, a freely available portable implementation of Message Passing Interface (MPI), is used to send and receive messages between the master computer and slave computers in this implementation. For more information about the Message Passing Interface, see [7].

It is important to assign equal amount of work to each slave computer. Two methods have been implemented for the assignment. One is dividing the interested space into a number of more or less equal regions. The other is random dividing the interested space into a number of regions according to the Latin squares method in [8]. The first method has a good performance when the number of slave computers is small. When the number of slave computers increases big enough, the latter method performs better.

The slave computers may complete their work allocations at different times. So the interactive distributed model (Fig. 1.4), which is developing in the implementation, is our next work. By doing this a slave computer can help others if it completes its work earlier. This will be quite helpful to enhance efficiency of the method.

1.3 The computation of the market split problem

1.3.1 Reformulation of the problem based on lattice basis reduction

The market split problem has been originally proposed as follows.

Definition 1. A large company has two divisions D_1 and D_2. The company supplies retailers with several products. The goal is to allocate each retailer to either division D_1 or division D_2 so that D_1 controls 40% of the company's market for each product and D_2 the remaining 60% or, if such a perfect 40/60 split is not possible for all the products, to minimize the sum of percentage deviations from the 40/60 split.

The market split problem has been formulated as a set of 0–1 linear programming instances in [9], and the feasibility version of the problem can be formulated mathematically as follows:

$$\exists \quad x_j \in \{0, 1\}, \quad j = 1, 2, \ldots, n,$$

$$\text{s.t.} \quad \sum_{j=1}^{n} c_{ij} x_j = d_i, i = 1, 2, \ldots, m? \tag{1.1}$$

Here m is the number of products, n is the number of retailers, c_{ij} is the demand of retailer j for product i, and the right-hand side vector d_i is demand from the desired market split among the two divisions D_1 and D_2. In [10], it is proved that the problem of this kind is NP-complete, and it is quite hard to solve it by traditional integer algorithms, especially by cutting plane method, branch-and-bound method, and relaxation techniques. Some excellent algorithms for (1.1) are provided in the literature [11–16]. However, these algorithms can just solve the market split problem with small dimension.

In this subsection, we introduce the outline algorithm in [13] for converting system (1.1) to a polytope judgment problem. The algorithm is based on the Lovász lattice basis reduction method described in [17]. Let N_1 and N_2 be positive integers that are large enough with respect to the input and in relation to each other. Consider the following linearly independent column vectors $\boldsymbol{B} = (\boldsymbol{b}_j)_{1 \le j \le n+1}$:

$$\boldsymbol{B} = \begin{pmatrix} \boldsymbol{e}^{(n)} & \boldsymbol{0}^{(n \times 1)} \\ \boldsymbol{0}^{(1 \times n)} & N_1 \\ N_2 \boldsymbol{C} & -N_2 \boldsymbol{d} \end{pmatrix},$$

where $\boldsymbol{C}$ denotes $(c_{ij})_{m \times n}$ in system (1.1), $\boldsymbol{d}$ denotes $(d_i)_m$ in system (1.1), $\boldsymbol{e}^{(n)}$ is the n-dimensional identity matrix, and $\boldsymbol{0}^{n \times 1}$ is the $n \times 1$-dimensional zero matrix. The LLL basis reduction algorithm, which is polynomial, is applied to the lattice $\boldsymbol{B}$, and then we obtain the following reformulation:

$$\hat{\boldsymbol{B}} = \begin{pmatrix} \boldsymbol{X}_0^{(n \times (n-m))} & \boldsymbol{x}_d \\ \boldsymbol{0}^{(1 \times (n-m))} & N_1 \\ \boldsymbol{0}^{(m \times (n-m))} & \boldsymbol{0}^{(m \times 1)} \end{pmatrix}.$$

Lemma 1 (Aardal et al. [13]). *Problem (1.1) can be formulated equivalently as follows:*
Does there exist a vector

$$\boldsymbol{\lambda} \in Z^{n-m} \quad \text{s.t.} - \boldsymbol{x}_d \le \boldsymbol{X}_0 \boldsymbol{\lambda} \le \boldsymbol{e}^{(n-m) \times 1} - \boldsymbol{x}_d? \tag{1.2}$$

Therefore we can judge whether there is an integer point in polytope (1.2) to solve problem (1.1) by the distributed implementation.

1.3.2 Numerical results

In this subsection, we present some numerical results to solve system (1.1). The input was generated as follows. For given $m = 5$, let $n = 10(m - 1) = 40$, and let the coefficients c_{ij} be integer numbers drawn uniformly and independently from the interval $[0, 99]$. The right-hand side coefficients are computed as $d_i = \lfloor \frac{1}{2} \sum_{j=1}^{n} c_{ij} \rfloor$, $1 \leq i \leq m$. Let $N_1 = 1000$ and $N_2 = 10\,000$ in the computation.

Given an example, two cases have been considered. The one is solving system (1.1) directly. The other is solving system (1.1) by computing system (1.2). Three other methods, including branch-and-cut method, branch-and-bound method, and Cplex constraint programming, have also been used to make comparison. The numerical results are presented in Tables 1.1 and 1.2. In the presentation of numerical results, we use the following symbols:

NumLPs: The number of iterations for a certain algorithm

F: Whether the example feasible or not

BC: The branch-and-cut method

BB: The branch-and-bound method

CP: Cplex constraint programming method

From Tables 1.1 and 1.2 we can see that this distributed method is the best one and the problem can be easier after the basis reduction. For a certain problem, the distributed method has the least number of iterations among these four algorithms. If the problem is solved without the basis reduction, then it is harder. For the branch-and-bound method, the problem of this dimension is beyond its ability.

1.4 The computation of the knapsack feasibility problem

The knapsack problems have been intensively studied since the pioneering work of [18], both because of their immediate applications in industry and financial management, but more pronounced for theoretical reasons, as knapsack problems frequently occur by relaxation of various integer programming problems. The knapsack feasibility problem is defined as follows.

Definition 2. Find a 0–1 integer solution of $p^T x = d$ such that $p = (p_1, p_2, \ldots, p_{n+1})^T > 0$ and $p_i \neq p_j$ for all $i \neq j$.

Table 1.1 Solving system (1.1) directly.

Prob.	BC		BB		CP	
	NumLPs	F	NumLPs	F	NumLPs	F
1	5.00E+07	Feasible	★★★	★★★	E+09	Feasible
2	6.70E+07	Feasible	★★★	★★★	E+09	Feasible
3	7.90E+07	Feasible	★★★	★★★	E+09	Feasible
4	5.30E+07	Feasible	★★★	★★★	2.59E+07	Feasible
5	2.60E+06	Feasible	★★★	★★★	1.24E+09	Feasible
6	2.60E+08	Infeasible	★★★	★★★	E+09	Infeasible
7	1.60E+08	Infeasible	★★★	★★★	E+09	Infeasible
8	8.20E+07	Infeasible	★★★	★★★	E+09	Infeasible
9	2.20E+08	Infeasible	★★★	★★★	E+09	Infeasible
10	1.10E+08	Infeasible	★★★	★★★	E+09	Infeasible
11	9.70E+07	Infeasible	★★★	★★★	E+09	Infeasible
12	4.50E+07	Feasible	★★★	★★★	E+09	Feasible
13	1.20E+08	Infeasible	★★★	★★★	E+09	Infeasible
14	1.10E+08	Infeasible	★★★	★★★	E+09	Infeasible
15	1.10E+08	Infeasible	★★★	★★★	E+09	Infeasible
16	1.60E+08	Feasible	★★★	★★★	E+09	Feasible
17	1.80E+07	Feasible	★★★	★★★	6.85E+08	Feasible
18	7.90E+07	Feasible	★★★	★★★	1.02E+09	Feasible
19	1.30E+08	Infeasible	★★★	★★★	E+09	Infeasible
20	1.40E+08	Infeasible	★★★	★★★	E+09	Infeasible
21	2.00E+08	Infeasible	★★★	★★★	E+09	Infeasible
22	2.90E+07	Feasible	★★★	★★★	E+09	Feasible
23	4.80E+07	Feasible	★★★	★★★	3.39E+08	Feasible
24	1.30E+08	Infeasible	★★★	★★★	E+09	Infeasible
25	2.40E+08	Infeasible	★★★	★★★	E+09	Infeasible

After analysis we can see that this knapsack problem is a special example of the market split problem. To convert this problem into an equivalent problem of determining whether there is an integer point in a full-dimensional polytope given by $P = \{x \in R^n | Ax \le b\}$, we can apply the basis reduction algorithm as described in Section 1.3.

Let $p_j \in [10^2, 10^4], j = 1, 2, \ldots, n+1$, $p_i \ne p_j$ for all $i \ne j$, let $d \in [10^2, 10^4]$ be generated randomly, and let $N_1 = 10\,000$ and $N_2 = 10\,000$. Some other parameters and symbols are the same as in Section 1.3. The numerical results are presented in Table 1.3.

From a number of iterations we can see that the distributed algorithm is superior to the branch-and-cut method.

Table 1.2 Solving system (1.2) after basis reduction.

Prob.	The method		BC		BB	
	NumLPs	F	NumLPs	F	NumLPs	F
1	9640	Feasible	4334	Feasible	1.50E+06	Feasible
2	32 015	Feasible	36 624	Feasible	1.2 E+06	Feasible
3	22 221	Feasible	23 652	Feasible	9.2 E+05	Feasible
4	12 670	Feasible	6924	Feasible	1.2 E+06	Feasible
5	49 709	Feasible	22 163	Feasible	2.5 E+06	Feasible
6	54 525	Infeasible	34 501	Infeasible	3.4 E+06	Infeasible
7	105 670	Infeasible	86 555	Infeasible	4.60E+06	Infeasible
8	90 204	Infeasible	77 661	Infeasible	5.00E+06	Infeasible
9	93 751	Infeasible	53 586	Infeasible	5.10E+06	Infeasible
10	67 565	Infeasible	51 837	Infeasible	3.3 E+06	Infeasible
11	90 218	Infeasible	58 848	Infeasible	7.20E+06	Infeasible
12	36 204	Feasible	43 794	Feasible	1.4 E+06	Feasible
13	106 082	Infeasible	111 431	Infeasible	3.90E+06	Infeasible
14	33 699	Infeasible	32 598	Infeasible	2.2 E+06	Infeasible
15	64 368	Infeasible	41 837	Infeasible	3.80E+06	Infeasible
16	38 577	Feasible	18 049	Feasible	1.9 E+06	Feasible
17	26 167	Feasible	7903	Feasible	1.2 E+06	Feasible
18	75 633	Feasible	4723	Feasible	4.40E+06	Feasible
19	86 061	Infeasible	48 813	Infeasible	4.50E+06	Infeasible
20	36 737	Infeasible	43 710	Infeasible	2.20E+06	Infeasible
21	67 556	Infeasible	46 083	Infeasible	5.70E+06	Infeasible
22	16 170	Feasible	58 411	Feasible	5.1 E+05	Feasible
23	33 848	Feasible	11 267	Feasible	2.20E+06	Feasible
24	78 172	Infeasible	40 769	Infeasible	3.70E+06	Infeasible
25	75 375	Infeasible	74 582	Infeasible	3.40E+06	Infeasible

1.5 Summary

In this chapter, we developed a distributed implementation of the Dang and Ye's fixed-point iterative method. Five sections are included in this chapter. Dang and Ye's fixed-point iterative method is introduced in the first section. Section 1.2 has presented the details of the distributed implementation. Two different kinds of problems, the computation of the market problem and the computation of the knapsack feasibility problem, have been tested by our distributed computing system. The numerical results are promising. This distributed implementation can be easily extended to computing other problems.

Table 1.3 The knapsack feasibility problems (0–1 solution).

Prob.	Dimension n	The method		The best CPLEX method	
		NumLPs	Feasibility	NumLPs	Feasibility
1	1000	1002	Feasible	3023	Feasible
2	1000	1010	Feasible	1542	Feasible
3	1000	1027	Feasible	1495	Feasible
4	1000	1011	Feasible	1428	Feasible
5	1000	1118	Feasible	883	Feasible
6	1000	1035	Feasible	2023	Feasible
7	1000	1000	Infeasible	1280	Infeasible
8	1000	1002	Feasible	1360	Feasible
9	1000	1002	Feasible	998	Feasible
10	1000	1013	Feasible	1087	Feasible
11	1000	1321	Feasible	1577	Feasible
12	1000	1003	Feasible	1117	Feasible
13	1000	1024	Feasible	1638	Feasible
14	1000	1005	Feasible	1122	Feasible
15	1000	1019	Feasible	1097	Feasible
16	1000	1007	Feasible	1365	Feasible
17	1000	999	Infeasible	1315	Infeasible
18	1000	1572	Feasible	3741	Feasible
19	1000	1031	Feasible	1170	Feasible
20	1000	1015	Feasible	2702	Feasible
21	1000	1002	Feasible	978	Feasible
22	1000	1007	Feasible	1486	Feasible
23	1000	1005	Feasible	1043	Feasible
24	1000	1001	Feasible	3643	Feasible
25	1000	1065	Feasible	2017	Feasible

References

[1] Chuangyin Dang, An increasing-mapping approach to integer programming based on lexicographic ordering and linear programming, in: Proceedings of the 9th International Symposium on Operations Research and Its Applications, 2010, pp. 55–60.

[2] Chuangyin Dang, Yinyu Ye, A fixed-point iterative approach to integer programming and distributed computation, TR-1, City University of Hong Kong, 2011.

[3] Zhengtian Wu, Chuangyin Dang, Changan Zhu, Searching one pure-strategy Nash equilibrium using a distributed computation approach, Journal Computers 9 (4) (2014) 859–866.

[4] George B. Dantzig, Linear Programming and Extensions, Princeton University Press, 1998.

[5] Yinyu Ye, Michael J. Todd, Shinji Mizuno, Mathematics of Operations Research (1994) 53–67.

[6] Yinyu Ye, Interior Point Algorithms: Theory and Analysis, vol. 44, John Wiley & Sons, 2011.

[7] Message Passing Interface Forum. MPI: a message-passing interface standard, version 2.2, http://www.mpiforum.org/docs/mpi-2.2/mpi22-report.pdf, 2009.

[8] Yuping Wang, Chuangyin Dang, An evolutionary algorithm for global optimization based on level-set evolution and Latin squares, IEEE Transactions on Evolutionary Computation 11 (5) (2007) 579–595.

[9] Gérard Cornuéjols, Milind Dawande, in: Integer Programming and Combinatorial Optimization, Springer, 1998, pp. 284–293.

[10] Michael R. Gary, David S. Johnson, Computers and Intractability: A Guide to the Theory of NP-Completeness, Freeman, San Francisco, 1979.

[11] Gérard Cornuéjols, Valid inequalities for mixed integer linear programs, Mathematical Programming 112 (1) (2008) 3–44.

[12] Karen Aardal, Cor Hurkens, Arjen K. Lenstra, Solving a linear Diophantine equation with lower and upper bounds on the variables, in: Integer Programming and Combinatorial Optimization, Springer, 1998, pp. 229–242.

[13] Karen Aardal, Robert E. Bixby, Cor A.J. Hurkens, Arjen K. Lenstra, Job W. Smeltink, Market split and basis reduction: Towards a solution of the Cornuéjols–Dawande instances, INFORMS Journal on Computing 12 (3) (2000) 192–202.

[14] Alfred Wassermann, Attacking the market split problem with lattice point enumeration, Journal of Combinatorial Optimization 6 (1) (2002) 5–16.

[15] Heiko Vogel, Solving market split problems with heuristical lattice reduction, Annals of Operations Research 196 (1) (2012) 581–590.

[16] M. Khorramizadeh, Numerical experiments with the lll-based Hermite normal form algorithm for solving linear Diophantine systems, International Journal of Contemporary Mathematical Sciences 7 (13) (2012) 599–613.

[17] Arjen Klaas Lenstra, Hendrik Willem Lenstra, László Lovász, Factoring polynomials with rational coefficients, Mathematische Annalen 261 (4) (1982) 515–534.

[18] George B. Dantzig, Discrete-variable extremum problems, Operations Research 5 (2) (1957) 266–288.

CHAPTER 2

Computing all pure-strategy Nash equilibria using mixed 0–1 linear programming approach

2.1 Converting the problem to a mixed 0–1 linear programming

Let $N = \{1, 2, \ldots, n\}$ be a set of players. The pure strategy set of player $i \in N$ is denoted by $S^i = \{s_j^i \mid j \in M_i\}$ with $M_i = \{1, 2, \ldots, m_i\}$. Given S^i with $i \in N$, the set of all pure strategy profiles is $S = \prod_{i=1}^n S^i$. We denote the payoff function of player $i \in N$ by $u^i : S \to R$. For $i \in N$, let $S^{-i} = \prod_{k \in N \setminus \{i\}} S^k$. Then $s = (s_{j_1}^1, s_{j_2}^2, \ldots, s_{j_n}^n) \in S$ can be rewritten as $s = (s_{j_i}^i, s^{-i})$ with $s^{-i} = (s_{j_1}^1, \ldots, s_{j_{i-1}}^{i-1}, s_{j_{i+1}}^{i+1}, \ldots, s_{j_n}^n) \in S^{-i}$. A mixed strategy of player i is a probability distribution on S^i denoted by $x^i = (x_1^i, x_2^i, \ldots, x_{m_i}^i)$. Let X^i be the set of all mixed strategies of player i. Then $X^i = \{x^i = (x_1^i, x_2^i, \ldots, x_{m_i}^i) \in R_+^{m_i} \mid \sum_{j=1}^{m_i} x_j^i = 1\}$. Thus, for $x^i \in X^i$, the probability assigned to pure strategy $s_j^i \in S^i$ is equal to x_j^i. Given X^i with $i \in N$, the set of all mixed strategy profiles is $X = \prod_{i=1}^n X^i$. For $i \in N$, let $X^{-i} = \prod_{k \in N \setminus \{i\}} X^k$. Then $x = (x^1, x^2, \ldots, x^n) \in X$ can be rewritten as $x = (x^i, x^{-i})$ with $x^{-i} = (x^1, \ldots, x^{i-1}, x^{i+1}, \ldots, x^n) \in X^{-i}$. If $x \in X$ is played, then the probability that a pure strategy profile $s = (s_{j_1}^1, s_{j_2}^2, \ldots, s_{j_n}^n) \in S$ occurs is $\prod_{i=1}^n x_{j_i}^i$. Therefore, for $x \in X$, the expected payoff of player i is given by $u^i(x) = \sum_{s \in S} u^i(s) \prod_{i=1}^n x_{j_i}^i$. With these notations, a finite n-person game in normal form can be represented as $\Gamma = \langle N, S, \{u^i\}_{i \in N} \rangle$ or $\Gamma = \langle N, X, \{u^i\}_{i \in N} \rangle$.

Definition 3 (Nash, [1]). A mixed strategy profile $x^* \in X$ is a Nash equilibrium of game Γ if $u^i(x^*) \geq u^i(x^i, x^{*-i})$ for all $i \in N$ and $x^i \in X^i$.

With this definition, an application of the optimality condition leads to that x^* is a Nash equilibrium if and only if there are λ^* and μ^* together with x^* satisfying the system

Integer Optimization and Its Computation in Emergency Management
https://doi.org/10.1016/B978-0-32-395203-3.00007-1

$$u^i(s^i_j, x^{-i}) + \lambda^i_j - \mu_i = 0,$$
$$e^{iT} x^i - 1 = 0,$$
$$x^i_j \lambda^i_j = 0, \tag{2.1}$$
$$x^i_j \geq 0, \ \lambda^i_j \geq 0,$$
$$j = 1, 2, \ldots, m_i, \ i = 1, 2, \ldots, n,$$

where $e^i = (1, 1, \ldots, 1)^\top \in R^{m_i}$.

Lemma 2. *Let β be a given positive number such that*

$$\beta \geq \max_{i \in N}\{\max_{s \in S} u^i(s) - \min_{s \in S} u^i(s)\}.$$

Then (2.1) is equivalent to

$$u^i(s^i_j, x^{-i}) + \lambda^i_j - \mu_i = 0,$$
$$e^{iT} x^i - 1 = 0,$$
$$x^i_j \leq v^i_j,$$
$$\lambda^i_j \leq \beta(1 - v^i_j), \tag{2.2}$$
$$v^i_j \in \{0, 1\},$$
$$x^i_j \geq 0, \ \lambda^i_j \geq 0,$$
$$j = 1, 2, \ldots, m_i, \ i = 1, 2, \ldots, n.$$

Proof. Let (x, λ, v) be a solution of system (2.2). Suppose that $v^i_j = 0$ in (2.2). Then $x^i_j = 0$ and $\lambda^i_j \leq \beta$. Thus $x^i_j \lambda^i_j = 0$. Suppose that $v^i_j = 1$ in (2.2). Then $\lambda^i_j = 0$. Thus $x^i_j \lambda^i_j = 0$. Therefore (x, λ) is a solution of system (2.1).

Let (x, λ) be a solution of system (2.1). Since $\sum_{j=1}^{m_i} x^i_j = 1$, there exists $k \in M_i$ such that $x^i_k > 0$. From $x^i_k \lambda^i_k = 0$ we get that $\lambda^i_k = 0$. Thus $\mu_i = u^i(s^i_k, x^{-i}) \leq \max_{s \in S} u^i(s)$. Therefore, for any $j \in M_i$, from system (2.1) we obtain that

$$\lambda^i_j = \mu_i - u^i(s^i_j, x^{-i}) \leq \max_{s \in S} u^i(s) - \min_{s \in S} u^i(s) \leq \beta.$$

Suppose that $x^i_j = 0$. Let $v^i_j = 0$. Thus

$$x^i_j = 0 \leq v^i_j,$$
$$\lambda^i_j \leq \beta(1 - v^i_j) = \beta.$$

Suppose $x^i_j > 0$. Let $v^i_j = 1$. Thus

$$x^i_j \leq v^i_j = 1,$$
$$\lambda^i_j = 0 \leq \beta(1 - v^i_j) = 0.$$

Therefore (x, λ, ν) is a solution of system (2.2). This completes the proof.

□

This lemma implies that finding a pure-strategy Nash equilibrium is equivalent to finding a solution of the system

$$
\begin{aligned}
&u^i(s^i_j, x^{-i}) + \lambda^i_j - \mu_i = 0, \\
&e^{iT} x^i - 1 = 0, \\
&\lambda^i_j \leq \beta(1 - x^i_j), \\
&x^i_j \in \{0, 1\}, \\
&\lambda^i_j \geq 0, \\
&j = 1, 2, \ldots, m_i,\ i = 1, 2, \ldots, n.
\end{aligned}
\tag{2.3}
$$

For any $s^i_j \in S^i$, from $u^i(x)$ we can obtain that

$$
u^i(s^i_j, x^{-i}) = \sum_{s^{-i}=(s^1_{j_1}, \ldots, s^{i-1}_{j_{i-1}}, s^{i+1}_{j_{i+1}}, \ldots, s^n_{j_n}) \in S^{-i}} u^i(s^i_j, s^{-i}) \prod_{k \neq i} x^k_{j_k}.
$$

Let $\gamma(s^{-i}) = \prod_{k \neq i} x^k_{j_k}$ for $s^{-i} = (s^1_{j_1}, \ldots, s^{i-1}_{j_{i-1}}, s^{i+1}_{j_{i+1}}, \ldots, s^n_{j_n}) \in S^{-i}$. Then

$$
u^i(s^i_j, x^{-i}) = \sum_{s^{-i}=(s^1_{j_1}, \ldots, s^{i-1}_{j_{i-1}}, s^{i+1}_{j_{i+1}}, \ldots, s^n_{j_n}) \in S^{-i}} u^i(s^i_j, s^{-i}) \gamma(s^{-i}).
$$

Substituting this into (2.3) yields

$$
\begin{aligned}
&\sum_{s^{-i} \in S^{-i}} u^i(s^i_j, s^{-i}) \gamma(s^{-i}) + \lambda^i_j - \mu_i = 0, \\
&e^{iT} x^i - 1 = 0, \\
&\lambda^i_j \leq \beta(1 - x^i_j), \\
&x^i_j \in \{0, 1\}, \\
&\lambda^i_j \geq 0, \\
&j = 1, 2, \ldots, m_i,\ i = 1, 2, \ldots, n, \\
&\gamma(s^{-i}) = \prod_{k \neq i} x^k_{j_k},\ s^{-i} \in S^{-i},\ i = 1, 2, \ldots, n.
\end{aligned}
\tag{2.4}
$$

Observe that $\gamma(s^{-i}) \in \{0, 1\}$ since $x^i_j \in \{0, 1\}$. This property leads to the following result.

Lemma 3. *For $i \in N$ and $s^{-i} = (s^1_{j_1}, \ldots, s^{i-1}_{j_{i-1}}, s^{i+1}_{j_{i+1}}, \ldots, s^n_{j_n}) \in S^{-i}$, the system*

$$
\begin{aligned}
&\gamma(s^{-i}) = \prod_{k \neq i} x^k_{j_k}, \\
&x^k_{j_k} \in \{0, 1\},\ k = 1, 2, \ldots, n,\ k \neq i,
\end{aligned}
\tag{2.5}
$$

is equivalent to the system

$$y(s^{-i}) \geq \sum_{h \neq i} x_{j_h}^h - (n-2),$$
$$y(s^{-i}) \leq x_{j_k}^k,$$
$$0 \leq y(s^{-i}),$$
$$x_{j_k}^k \in \{0, 1\}, \ k = 1, 2, \ldots, n, \ k \neq i. \tag{2.6}$$

Proof. Let (y, x) be a solution of system (2.6). Suppose that $x_{j_k}^k = 1$ for all $k \neq i$. Then

$$y(s^{-i}) \geq \sum_{h \neq i} x_{j_h}^h - (n-2) = (n-1) - (n-2) = 1,$$
$$y(s^{-i}) \leq x_{j_k}^k = 1, \ k \neq i.$$

Thus $y(s^{-i}) = \prod_{k \neq i} x_{j_k}^k = 1$.

Suppose that there is $k \neq i$ such that $x_{j_k}^k = 0$. Then

$$\exists k \neq i \text{ s.t. } y(s^{-i}) \leq x_{j_k}^k = 0,$$
$$0 \leq y(s^{-i}).$$

Thus $y(s^{-i}) = \prod_{k \neq i} x_{j_k}^k = 0$. Therefore (y, x) is a solution of system (2.5).

Let (y, x) be a solution of system (2.5). Suppose that $x_{j_k}^k = 1$ for all $k \neq i$. Then $y(s^{-i}) = \prod_{k \neq i} x_{j_k}^k = 1$. Thus

$$\sum_{h \neq i} x_{j_h}^h - (n-2) = (n-1) - (n-2) = 1 \leq y(s^{-i}) = 1,$$
$$y(s^{-i}) = 1 \leq x_{j_k}^k = 1, \ k \neq i,$$
$$0 \leq y(s^{-i}).$$

Suppose that there is $k \neq i$ such that $x_{j_k}^k = 0$. Then $y(s^{-i}) = \prod_{k \neq i} x_{j_k}^k = 0$. Thus

$$\sum_{h \neq i} x_{j_h}^h - (n-2) \leq (n-2) - (n-2) = y(s^{-i}) = 0,$$
$$y(s^{-i}) = 0 \leq x_{j_k}^k, \ k \neq i,$$
$$0 \leq y(s^{-i}).$$

Therefore (y, x) is a solution of system (2.6). This completes the proof. $\square$

Replacing $\gamma(s^{-i}) = \prod_{k \neq i} x_{j_k}^k$ of system (2.4) with system (2.6), we obtain the mixed 0–1 linear program

$$
\begin{aligned}
&\sum_{s^{-i} \in S^{-i}} u^i(s_j^i, s^{-i})\gamma(s^{-i}) + \lambda_j^i - \mu_i = 0, \\
&e^{iT} x^i - 1 = 0, \\
&\lambda_j^i \leq \beta(1 - x_j^i), \\
&x_j^i \in \{0, 1\}, \\
&\lambda_j^i \geq 0, \\
&j = 1, 2, \ldots, m_i, \ i = 1, 2, \ldots, n, \\
&\gamma(s^{-i}) \geq \sum_{h \neq i} x_{j_h}^h - (n - 2), \\
&\gamma(s^{-i}) \leq x_{j_k}^k, \ k \neq i, \\
&0 \leq \gamma(s^{-i}), \\
&s^{-i} \in S^{-i}, \ i = 1, 2, \ldots, n.
\end{aligned} \tag{2.7}
$$

As a corollary of Lemma 3, we obtain the following result.

Corollary 1. *System (2.7) is equivalent to system (2.3).*

The above idea is illustrated by the following example.

Example 1. Consider a three-player game $\Gamma = (N, S, \{u^i\}_{i \in N})$, where $N = \{1, 2, 3\}$, $S^i = \{s_1^i, s_2^i\}$, $i \in N$, and $\{u^i\}_{i \in N}$ are given by

	s_1^2	s_2^2	s_1^2	s_2^2
s_1^1	(1,1,1)	(1,0,1)	(0,1,0)	(1,0,0)
s_2^1	(1,1,1)	(1,0,1)	(0,1,0)	(0,0,0)
	s_1^3		s_2^3	

Let $\beta = 1$. The mixed 0–1 linear program for this game is given by

$$
\begin{aligned}
&\gamma(s_1^2, s_1^3) + \gamma(s_2^2, s_1^3) + \gamma(s_2^2, s_2^3) + \lambda_1^1 - \mu_1 = 0, \\
&\gamma(s_1^2, s_1^3) + \gamma(s_2^2, s_1^3) + \lambda_2^1 - \mu_1 = 0, \\
&x_1^1 + x_2^1 = 1, \\
&\gamma(s_1^1, s_1^3) + \gamma(s_2^1, s_1^3) + \gamma(s_1^1, s_2^3) + \gamma(s_2^1, s_2^3) + \lambda_1^2 - \mu_2 = 0, \\
&\lambda_2^2 - \mu_2 = 0, \\
&x_1^2 + x_2^2 = 1, \\
&\gamma(s_1^1, s_1^2) + \gamma(s_2^1, s_1^2) + \gamma(s_1^1, s_2^2) + \gamma(s_2^1, s_2^2) + \lambda_1^3 - \mu_3 = 0, \\
&\lambda_2^3 - \mu_3 = 0, \\
&x_1^3 + x_2^3 = 1,
\end{aligned}
$$

$$\lambda_j^i \leq 1 - x_j^i, \quad x_j^i \in \{0, 1\},$$

$$\lambda_j^i \geq 0,$$

$$j = 1, 2,$$

$$i = 1, 2, 3,$$

$$\gamma(s_1^2, s_1^3) \geq x_1^2 + x_1^3 - 1,$$

$$\gamma(s_1^2, s_1^3) \leq x_1^2,$$

$$\gamma(s_1^2, s_1^3) \leq x_1^3,$$

$$\gamma(s_1^2, s_1^3) \geq 0,$$

$$\gamma(s_1^2, s_2^3) \geq x_1^2 + x_2^3 - 1,$$

$$\gamma(s_1^2, s_2^3) \leq x_1^2,$$

$$\gamma(s_1^2, s_2^3) \leq x_2^3,$$

$$\gamma(s_1^2, s_2^3) \geq 0,$$

$$\gamma(s_2^2, s_1^3) \geq x_2^2 + x_1^3 - 1,$$

$$\gamma(s_2^2, s_1^3) \leq x_2^2,$$

$$\gamma(s_2^2, s_1^3) \leq x_1^3,$$

$$\gamma(s_2^2, s_1^3) \geq 0,$$

$$\gamma(s_2^2, s_2^3) \geq x_2^2 + x_2^3 - 1,$$

$$\gamma(s_2^2, s_2^3) \leq x_2^2,$$

$$\gamma(s_2^2, s_2^3) \leq x_2^3,$$

$$\gamma(s_2^2, s_2^3) \geq 0,$$

$$\gamma(s_1^1, s_1^3) \geq x_1^1 + x_1^3 - 1,$$

$$\gamma(s_1^1, s_1^3) \leq x_1^1,$$

$$\gamma(s_1^1, s_1^3) \leq x_1^3,$$

$$\gamma(s_1^1, s_1^3) \geq 0,$$

$$\gamma(s_2^1, s_1^3) \geq x_2^1 + x_1^3 - 1,$$

$$\gamma(s_2^1, s_1^3) \leq x_2^1,$$

$$\gamma(s_2^1, s_1^3) \leq x_1^3,$$

$$\gamma(s_2^1, s_1^3) \geq 0,$$

$$\gamma(s_1^1, s_2^3) \geq x_1^1 + x_2^3 - 1,$$

$$\gamma(s_1^1, s_2^3) \leq x_1^1,$$

$$\gamma(s_1^1, s_2^3) \leq x_2^3,$$

$$\gamma(s_1^1, s_2^3) \geq 0,$$

$$\gamma(s_2^1, s_2^3) \geq x_2^1 + x_2^3 - 1,$$

$$\gamma(s_2^1, s_2^3) \leq x_2^1,$$

$$\gamma(s_2^1, s_2^3) \leq x_2^3,$$

$$\gamma(s_2^1, s_2^3) \geq 0,$$

$$\gamma(s_1^1, s_1^2) \geq x_1^1 + x_1^2 - 1,$$

$$\gamma(s_1^1, s_1^2) \leq x_1^1,$$

$$\gamma(s_1^1, s_1^2) \leq x_1^2,$$

$$\gamma(s_1^1, s_1^2) \geq 0,$$

$$\gamma(s_1^1, s_2^2) \geq x_1^1 + x_2^2 - 1,$$

$$\gamma(s_1^1, s_2^2) \leq x_1^1,$$

$$\gamma(s_1^1, s_2^2) \leq x_2^2,$$

$$\gamma(s_1^1, s_2^2) \geq 0,$$

$$\gamma(s_2^1, s_1^2) \geq x_2^1 + x_1^2 - 1,$$

$$\gamma(s_2^1, s_1^2) \leq x_2^1,$$

$$\gamma(s_2^1, s_1^2) \leq x_1^2,$$

$$\gamma(s_2^1, s_1^2) \geq 0,$$

$$\gamma(s_2^1, s_2^2) \geq x_2^1 + x_2^2 - 1,$$

$$\gamma(s_2^1, s_2^2) \leq x_2^1,$$

$$\gamma(s_2^1, s_2^2) \leq x_2^2,$$

$$\gamma(s_2^1, s_2^2) \geq 0.$$

2.2 Numerical results

In this section, we present some numerical results. We use C++ to call ILOG CPLEX API functions to solve the mixed 0–1 linear program (2.7). The MIP (mixed integer programming) search method, which is a dynamic search strategy or branch–and–cut strategy in ILOG CPLEX, is automatically determined by the ILOG CPLEX. All other parameters are also automatically set by ILOG CPLEX itself.

We have run our code on a workstation of Lenovo ThinkStation D20 4155-BM4 with 16 processors and 16 G RAM. In the presentation of numerical results, we use the following symbols.

NumN: The number of players.

NumS: The number of strategies for each player.

NumEquilibra: The number of pure-strategy Nash equilibria for the instance.

Time: The total computational time to solve the problem.

Example 2. Consider a three-player game $\Gamma = (N, S, \{u^i\}_{i \in N})$, where $N = \{1, 2, 3\}$. The number of strategies for each player, *NumS*, is randomly generated from 3 to 45. The $\{u^i\}_{i \in N}$ are also randomly generated. There are three different ranges for $\{u^i\}_{i \in N}$ in this example, which are from 0 to 10, from 0 to 50, and from 0 to 100.

Let $\beta = 1000$; 30 instances have been solved for each range of $\{u^i\}_{i \in N}$. Numerical results for each range are presented in Tables 2.1, 2.2, and 2.3, respectively.

Although every n-player game can be reduced to a three-player game in polynomial time as shown in [2]. Some five-player games are also solved to test the efficiency of this mixed 0–1 linear programming approach.

Example 3. Consider a five-player game $\Gamma = (N, S, \{u^i\}_{i \in N})$, where $N = \{1, 2, 3, 4, 5\}$. The number of strategies for each player, *NumS*, is randomly generated from 3 to 7. The $\{u^i\}_{i \in N}$ are also randomly generated. There are three different ranges for $\{u^i\}_{i \in N}$ in this example, which are from 0 to 10, from 0 to 50, and from 0 to 100.

Let $\beta = 1000$; 10 instances have been solved for each range of $\{u^i\}_{i \in N}$. Numerical results for each range are presented in Tables 2.4, 2.5, and 2.6, respectively.

From Tables 2.1, 2.2, and 2.3 we can see that this mixed 0–1 linear programming method can effectively find all pure-strategy Nash equilibria of the three-person game. For example, in Table 2.1, problem 30 with 45 strategies can be solved in 16992.10 seconds, and the number of equilibria is 26. It has been showed that every n-person game can be reduced to a three-person game in polynomial time [2]. However, some five-person

Table 2.1 Three-player game with $u^i(s^i_j, s^{-i})$ randomly generated from 0 to 10.

Prob.	NumN	NumS	NumEquilibra	Time (s)
1	3	4	1	0.53
2	3	4	2	0.62
3	3	4	2	0.56
4	3	5	3	0.56
5	3	6	2	0.41
6	3	6	1	0.66
7	3	7	5	0.59
8	3	8	6	0.58
9	3	8	3	0.89
10	3	8	3	0.81
11	3	10	2	1.00
12	3	11	3	1.47
13	3	12	4	1.81
14	3	14	4	4.91
15	3	14	8	3.81
16	3	15	6	5.87
17	3	19	13	27.86
18	3	19	3	33.40
19	3	21	13	47.28
20	3	23	10	102.51
21	3	24	22	122.94
22	3	25	21	239.20
23	3	26	13	278.50
24	3	26	15	301.58
25	3	28	21	419.97
26	3	33	22	1870.24
27	3	35	20	2082.34
28	3	40	24	4599.72
29	3	44	20	3033.29
30	3	45	26	16 992.10

games are still solved by this method in our examples to test the effectiveness of this mixed 0–1 linear programming approach. The numerical results for a five-person game are presented in Tables 2.4, 2.5, and 2.6. All the numerical results show that this new method can successfully handle all pure-strategy Nash equilibria of an n-person game. However, the computa-

Table 2.2 Three-player game with $u^i(s^i_j, s^{-i})$ randomly generated from 0 to 50.

Prob.	NumN	NumS	NumEquilibra	Time (s)
1	3	3	4	0.46
2	3	3	0	0.50
3	3	3	1	0.40
4	3	4	0	0.34
5	3	4	2	0.52
6	3	6	0	0.45
7	3	7	1	0.70
8	3	7	2	0.82
9	3	7	1	0.57
10	3	8	0	0.73
11	3	9	0	0.66
12	3	10	0	0.93
13	3	11	1	1.64
14	3	16	0	13.26
15	3	19	0	33.22
16	3	20	3	40.67
17	3	21	2	55.10
18	3	22	4	74.20
19	3	24	0	195.78
20	3	24	2	142.14
21	3	27	3	507.44
22	3	29	2	1178.18
23	3	30	1	1287.87
24	3	30	1	1667.70
25	3	32	1	1638.94
26	3	33	2	1761.86
27	3	34	0	3125.08
28	3	35	5	3446.67
29	3	44	3	36 601.30
30	3	45	5	43 583.30

tion time increases quickly with the increase of the problem dimension. To tackle this problem, the distributed computation, which is a salient feature of the integer programming, will be considered in our next work. Besides, some similar problems with multilinear terms can be also solved by this method.

Table 2.3 Three-player game with $u^i(s^i_j, s^{-i})$ randomly generated from 0 to 100.

Prob.	NumN	NumS	NumEquilibra	Time (s)
1	3	3	0	0.41
2	3	3	2	0.45
3	3	4	2	0.51
4	3	5	0	0.49
5	3	5	2	0.61
6	3	7	0	0.56
7	3	7	2	0.41
8	3	8	0	0.61
9	3	8	2	0.89
10	3	8	2	0.68
11	3	10	1	1.11
12	3	12	1	4.00
13	3	13	1	2.64
14	3	13	2	2.64
15	3	16	0	14.25
16	3	19	4	25.70
17	3	22	4	55.94
18	3	24	2	132.87
19	3	24	1	142.96
20	3	25	2	246.76
21	3	26	2	499.40
22	3	28	0	804.76
23	3	28	1	671.67
24	3	30	2	1382.03
25	3	32	1	2821.16
26	3	34	2	3325.67
27	3	37	1	7088.06
28	3	38	2	9395.13
29	3	41	2	20 441.10
30	3	43	0	29 204.70

2.3 Summary

In this chapter, a mixed 0–1 linear programming has been formulated to find all pure-strategy Nash equilibria of a finite game in normal form, and some numerical results of this computing method have been given.

Table 2.4 Five-player game with $u^i(s^i_j, s^{-i})$ randomly generated from 0 to 10.

Prob.	NumN	NumS	NumEquilibra	Time (s)
1	5	3	4	1.09
2	5	3	3	1.18
3	5	3	2	1.19
4	5	4	3	6.85
5	5	4	4	5.26
6	5	5	2	98.95
7	5	5	3	152.76
8	5	5	1	122.88
9	5	6	4	2092.13
10	5	6	2	1693.61

Table 2.5 Five-player game with $u^i(s^i_j, s^{-i})$ randomly generated from 0 to 50.

Prob.	NumN	NumS	NumEquilibra	Time(s)
1	5	3	3	1.20
2	5	3	1	0.99
3	5	3	1	1.71
4	5	4	1	5.69
5	5	4	1	5.62
6	5	4	1	11.90
7	5	4	3	6.77
8	5	5	1	1996.60
9	5	6	2	2031.19
10	5	7	3	18094.80

Table 2.6 Five-player game with $u^i(s^i_j, s^{-i})$ randomly generated from 0 to 100.

Prob.	NumN	NumS	NumEquilibra	Time (s)
1	5	3	0	0.80
2	5	4	0	10.98
3	5	4	0	11.11
4	5	4	1	6.93
5	5	4	0	11.07
6	5	5	1	99.30
7	5	5	0	138.33
8	5	6	1	2408.44
9	5	6	2	1960.49
10	5	7	3	22237.60

In this section, the formal definition of a Nash equilibrium is first introduced. With this definition, an application of the optimality condition leads to that a problem has a Nash equilibrium if and only if system (2.1) has a solution. After exploiting the properties of pure strategy and multilinear terms in the payoff functions, an equivalent formulation (2.7) of system (2.1) has been developed. We can solve system (2.7) to obtain all pure-strategy Nash equilibria. The example given in the first section explains the computing process. Some numerical results of computing Nash equilibria presented in the second section are promising. More details about the method can be found in [3].

References

[1] John F. Nash, Non-cooperative games, Annals of Mathematics 54 (2) (1951) 286–295.
[2] V. Bubelis, On equilibria in finite games, International Journal of Game Theory 8 (2) (1979) 65–79.
[3] Zhengtian Wu, Chuangyin Dang, Hamid Reza Karimi, Changan Zhu, Qing Gao, A mixed 0–1 linear programming approach to the computation of all pure-strategy Nash equilibria of a finite n-person game in normal form, Mathematical Problems in Engineering (2014) 2014.

CHAPTER 3

Computing all mixed-strategy Nash equilibria using mixed integer linear programming approach

3.1 Converting the problem to a mixed integer linear programming

With the formal definition of Nash equilibrium 3 in Chapter 2, an application of the optimality condition leads to that x^* is a Nash equilibrium if and only if there are λ^* and μ^* together with x^* satisfying the system

$$
\begin{aligned}
&u^i(s^i_j, x^{-i}) + \lambda^i_j - \mu_i = 0, \\
&e^{i\top} x^i - 1 = 0, \\
&x^i_j \lambda^i_j = 0, \\
&x^i_j \geq 0, \ \lambda^i_j \geq 0, \\
&j = 1, 2, \ldots, m_i, \ i = 1, 2, \ldots, n,
\end{aligned}
\tag{3.1}
$$

where $e^i = (1, 1, \ldots, 1)^\top \in R^{m_i}$.

Let β be a given positive number such that

$$
\beta \geq \max_{i \in N} \left\{ \max_{s \in S} u^i(s) - \min_{s \in S} u^i(s) \right\}.
$$

Then Lemma 2 in Chapter 2 shows that system (3.1) is equivalent to the following system:

$$
\begin{aligned}
&u^i(s^i_j, x^{-i}) + \lambda^i_j - \mu_i = 0, \\
&e^{i\top} x^i - 1 = 0, \\
&x^i_j \leq v^i_j, \\
&\lambda^i_j \leq \beta(1 - v^i_j), \\
&v^i_j \in \{0, 1\}, \\
&x^i_j \geq 0, \ \lambda^i_j \geq 0, \\
&j = 1, 2, \ldots, m_i, \ i = 1, 2, \ldots, n.
\end{aligned}
\tag{3.2}
$$

Integer Optimization and Its Computation in Emergency Management
https://doi.org/10.1016/B978-0-32-395203-3.00008-3

For any $s_j^i \in S^i$, from $u^i(x)$ we can obtain that

$$u^i\left(s_j^i, x^{-i}\right) = \sum_{s^{-i}=(s_{j_1}^1, \ldots, s_{j_{i-1}}^{i-1}, s_{j_{i+1}}^{i+1}, \ldots, s_{j_n}^n) \in S^{-i}} u^i\left(s_j^i, s^{-i}\right) \prod_{k \neq i} x_{j_k}^k.$$

Let $q(s^{-i}) = \prod_{k \neq i} x_{j_k}^k$ for $s^{-i} = (s_{j_1}^1, \ldots, s_{j_{i-1}}^{i-1}, s_{j_{i+1}}^{i+1}, \ldots, s_{j_n}^n) \in S^{-i}$. Then

$$u^i\left(s_j^i, x^{-i}\right) = \sum_{s^{-i}=(s_{j_1}^1, \ldots, s_{j_{i-1}}^{i-1}, s_{j_{i+1}}^{i+1}, \ldots, s_{j_n}^n) \in S^{-i}} u^i\left(s_j^i, s^{-i}\right) q\left(s^{-i}\right).$$

Substituting this into (3.2) yields

$$
\begin{aligned}
&\sum_{s^{-i} \in S^{-i}} u^i(s_j^i, s^{-i}) q(s^{-i}) + \lambda_j^i - \mu_i = 0, \\
&e^{iT} x^i - 1 = 0, \\
&x_j^i \leq v_j^i, \\
&\lambda_j^i \leq \beta(1 - v_j^i), \\
&v_j^i \in \{0, 1\}, \\
&x_j^i \geq 0, \ \lambda_j^i \geq 0, \\
&j = 1, 2, \ldots, m_i, \ i = 1, 2, \ldots, n, \\
&q(s^{-i}) = \prod_{k \neq i} x_{j_k}^k, \ s^{-i} \in S^{-i}, \ i = 1, 2, \ldots, n.
\end{aligned}
\tag{3.3}
$$

For $i \in N$ and $s^{-i} = (s_{j_1}^1, \ldots, s_{j_{i-1}}^{i-1}, s_{j_{i+1}}^{i+1}, \ldots, s_{j_n}^n) \in S^{-i}$, let

$$
\begin{aligned}
&\gamma_0(s^{-i}) = 1, \\
&\gamma_k(s^{-i}) = x_{j_k}^k \gamma_{k-1}(s^{-i}), \ k = 1, 2, \ldots, i-1, \\
&\gamma_{i+1}(s^{-i}) = x_{j_{i+1}}^{i+1} \gamma_{i-1}(s^{-i}), \\
&\gamma_k(s^{-i}) = x_{j_k}^k \gamma_{k-1}(s^{-i}), \ k = i+2, i+3, \ldots, n.
\end{aligned}
\tag{3.4}
$$

Then

$$
q\left(s^{-i}\right) = \begin{cases} \gamma_n(s^{-i}) & \text{if } i < n, \\ \gamma_{n-1}(s^{-i}) & \text{if } i = n. \end{cases}
$$

Therefore we can solve system (3.3) to find all Nash equilibria of system (3.1).

The paper [1] studies approaches for obtaining convex relaxation of global optimization problems containing multilinear functions. Consider $y = x_1 x_2$ with $x_1 \in [l_1, u_1]$ and $x_2 \in [l_2, u_2]$, where $u_1 > l_1 \geq 0$ and $u_2 > l_2 \geq 0$.

Lemma 4. *Let*

$$D = \left\{ (x_1, x_2, y)^\top \;\middle|\; \begin{array}{l} y \geq l_2 x_1 + l_1 x_2 - l_1 l_2, \\ y \geq u_2 x_1 + u_1 x_2 - u_1 u_2, \\ y \leq l_2 x_1 + u_1 x_2 - l_2 u_1, \\ y \leq u_2 x_1 + l_1 x_2 - l_1 u_2 \end{array} \right\}.$$

Then D is the convex hull of

$$\left\{ (x_1, x_2, y)^\top \;\middle|\; y = x_1 x_2,\; l_1 \leq x_1 \leq u_1,\; l_2 \leq x_2 \leq u_2 \right\}.$$

The proof of Lemma 4 can be found in [2].

Let d be any given positive integer. For $h = 1, 2, \ldots, d$, let $z_h \in \{0, 1\}$ be such that

$$(1 - z_h) l_2 + \left\{ l_2 + \frac{h-1}{d}(u_2 - l_2) \right\} z_h \leq x_2 \leq z_h \left\{ l_2 + \frac{h}{d}(u_2 - l_2) \right\} + (1 - z_h) u_2.$$

Then, for any $x_1 \in [l_1, u_1]$ and $x_2 \in [l_2, u_2]$, y satisfying

$$y \geq \left\{ l_2 + \frac{h-1}{d}(u_2 - l_2) \right\} x_1 + l_1 x_2 - l_1 \left\{ l_2 + \frac{h-1}{d}(u_2 - l_2) \right\}$$
$$- (1 - z_h)(u_1 u_2 - l_1 l_2),$$

$$y \geq \left\{ l_2 + \frac{h}{d}(u_2 - l_2) \right\} x_1 + u_1 x_2 - u_1 \left\{ l_2 + \frac{h}{d}(u_2 - l_2) \right\} - (1 - z_h)(u_1 u_2 - l_1 l_2),$$

$$y \leq \left\{ l_2 + \frac{h-1}{d}(u_2 - l_2) \right\} x_1 + u_1 x_2 - \left\{ l_2 + \frac{h-1}{d}(u_2 - l_2) \right\} u_1$$
$$+ (1 - z_h)(u_1 u_2 - l_1 l_2),$$

$$y \leq \left\{ l_2 + \frac{h}{d}(u_2 - l_2) \right\} x_1 + l_1 x_2 - l_1 \left\{ l_2 + \frac{h}{d}(u_2 - l_2) \right\} + (1 - z_h)(u_1 u_2 - l_1 l_2),$$

$$(1 - z_h) l_2 + \left\{ l_2 + \frac{h-1}{d}(u_2 - l_2) \right\} z_h \leq x_2 \leq z_h \left\{ l_2 + \frac{h}{d}(u_2 - l_2) \right\} + (1 - z_h) u_2,$$

$$z_h \in \{0, 1\}, \; h = 1, 2, \ldots, d,$$

$$\sum_{j=1}^{d} z_j = 1,$$

yields a good approximation of $x_1 x_2$.

For $h = 1, 2, \ldots, d$ and $x_{j_k}^k \in [\frac{h-1}{d}, \frac{h}{d}]$, we easily see that

$$0 \leq y_k(s^{-i}) = y_{k-1}(s^{-i}) x_{j_k}^k \leq \frac{h}{d}.$$

For $h = 1, 2, \ldots, d$, let $z_h^{kj_k} \in \{0, 1\}$ be such that

$$\frac{h-1}{d} z_h^{kj_k} \leq x_{j_k}^k \leq z_h^{kj_k} \frac{h}{d} + 1 - z_h^{kj_k}.$$

Then any point in

$$\left\{ \left(y_{k-1}(s^{-i}), x_{j_k}^k, y_k(s^{-i}) \right)^{\top} \middle| \begin{array}{l} y_k(s^{-i}) \geq \frac{h-1}{d} y_{k-1}(s^{-i}) - (1 - z_h^{kj_k}), \\ y_k(s^{-i}) \geq \frac{h}{d} y_{k-1}(s^{-i}) + x_{j_k}^k - \frac{h}{d} - (1 - z_h^{kj_k}), \\ y_k(s^{-i}) \leq \frac{h-1}{d} y_{k-1}(s^{-i}) + x_{j_k}^k - \frac{h-1}{d} + (1 - z_h^{kj_k}), \\ y_k(s^{-i}) \leq \frac{h}{d} y_{k-1}(s^{-i}) + (1 - z_h^{kj_k}), \\ \frac{h-1}{d} z_h^{kj_k} \leq x_{j_k}^k \leq z_h^{kj_k} \frac{h}{d} + 1 - z_h^{kj_k}, \\ z_h^{kj_k} \in \{0, 1\}, \\ h = 1, 2, \ldots, d, \\ \sum_{h=1}^{d} z_h^{kj_k} = 1 \end{array} \right\}$$

yields a good approximation of $(y_{k-1}(s^{-i}), x_{j_k}^k, y_k(s^{-i}))^{\top}$ with $y_k(s^{-i}) = y_{k-1}(s^{-i}) x_{j_k}^k$, $y_{k-1}(s^{-i}) \in [0, 1]$, and $x_{j_k}^k \in [0, 1]$. This, together with (3.4), leads to

$$
\begin{aligned}
& y_0(s^{-i}) = 1, \\
& y_k(s^{-i}) \geq \frac{h-1}{d} y_{k-1}(s^{-i}) - (1 - z_h^{kj_k}), \\
& y_k(s^{-i}) \geq \frac{h}{d} y_{k-1}(s^{-i}) + x_{j_k}^k - \frac{h}{d} - (1 - z_h^{kj_k}), \\
& y_k(s^{-i}) \leq \frac{h-1}{d} y_{k-1}(s^{-i}) + x_{j_k}^k - \frac{h-1}{d} + (1 - z_h^{kj_k}), \\
& y_k(s^{-i}) \leq \frac{h}{d} y_{k-1}(s^{-i}) + (1 - z_h^{kj_k}), \\
& \frac{h-1}{d} z_h^{kj_k} \leq x_{j_k}^k \leq z_h^{kj_k} \frac{h}{d} + 1 - z_h^{kj_k}, \\
& z_h^{kj_k} \in \{0, 1\}, \qquad\qquad\qquad\qquad\qquad\qquad\qquad (3.5) \\
& h = 1, 2, \ldots, d, \\
& \sum_{h=1}^{d} z_h^{kj_k} = 1, \\
& k = 1, 2, \ldots, i - 1, \ y_{i+1}(s^{-i}) \geq \frac{h-1}{d} y_{i-1}(s^{-i}) - (1 - z_h^{i+1,j_{i+1}}), \\
& y_{i+1}(s^{-i}) \geq \frac{h}{d} y_{i-1}(s^{-i}) + x_{j_{i+1}}^{i+1} - \frac{h}{d} - (1 - z_h^{i+1,j_{i+1}}), \\
& y_{i+1}(s^{-i}) \leq \frac{h-1}{d} y_{i-1}(s^{-i}) + x_{j_{i+1}}^{i+1} - \frac{h-1}{d} + (1 - z_h^{i+1,j_{i+1}}), \\
& y_{i+1}(s^{-i}) \leq \frac{h}{d} y_{i-1}(s^{-i}) + (1 - z_h^{i+1,j_{i+1}}),
\end{aligned}
$$

$$\frac{h-1}{d}z_h^{i+1,j_{i+1}} \le x_{j_{i+1}}^{i+1} \le z_h^{i+1,j_{i+1}}\frac{h}{d} + 1 - z_h^{i+1,j_{i+1}},$$

$$z_h^{i+1,j_{i+1}} \in \{0,1\},$$

$$h = 1,2,\ldots,d,$$

$$\sum_{h=1}^{d} z_h^{i+1,j_{i+1}} = 1,$$

$$y_k(s^{-i}) \ge \frac{h-1}{d}y_{k-1}(s^{-i}) - (1 - z_h^{kj_k}),$$

$$y_k(s^{-i}) \ge \frac{h}{d}y_{k-1}(s^{-i}) + x_{j_k}^{k} - \frac{h}{d} - (1 - z_h^{kj_k}),$$

$$y_k(s^{-i}) \le \frac{h-1}{d}y_{k-1}(s^{-i}) + x_{j_k}^{k} - \frac{h-1}{d} + (1 - z_h^{kj_k}), \qquad (3.6)$$

$$y_k(s^{-i}) \le \frac{h}{d}y_{k-1}(s^{-i}) + (1 - z_h^{kj_k}),$$

$$\frac{h-1}{d}z_h^{kj_k} \le x_{j_k}^{k} \le z_h^{kj_k}\frac{h}{d} + 1 - z_h^{kj_k},$$

$$z_h^{kj_k} \in \{0,1\},$$

$$h = 1,2,\ldots,d,$$

$$\sum_{h=1}^{d} z_h^{kj_k} = 1,$$

$$k = i+2, i+3, \ldots, n.$$

Replacing $q(s^{-i})$ of system (3.3) with systems (3.5) and (3.6), we obtain the following mixed–integer linear programming:

$$\sum_{s^{-i} \in S^{-i}} u^i(s_j^i, s^{-i})q(s^{-i}) + \lambda_j^i - \mu_i = 0,$$

$$e^{iT}x^i - 1 = 0,$$

$$x_j^i \le v_j^i,$$

$$\lambda_j^i \le \beta(1 - v_j^i),$$

$$v_j^i \in \{0,1\},$$

$$x_j^i \ge 0,$$

$$\lambda_j^i \ge 0,$$

$$j = 1,2,\ldots,m_i, \ i = 1,2,\ldots,n,$$

$$y_0(s^{-i}) = 1,$$

$$y_k(s^{-i}) \ge \frac{h-1}{d}y_{k-1}(s^{-i}) - (1 - z_h^{kj_k}),$$

$$y_k(s^{-i}) \ge \frac{h}{d}y_{k-1}(s^{-i}) + x_{j_k}^{k} - \frac{h}{d} - (1 - z_h^{kj_k}),$$

$$y_k(s^{-i}) \le \frac{h-1}{d}y_{k-1}(s^{-i}) + x_{j_k}^{k} - \frac{h-1}{d} + (1 - z_h^{kj_k}),$$

$$y_k(s^{-i}) \le \frac{h}{d}y_{k-1}(s^{-i}) + (1 - z_h^{kj_k}),$$

$$\frac{h-1}{d}z_h^{kj_k} \leq x_{j_k}^k \leq z_h^{kj_k}\frac{h}{d}+1-z_h^{kj_k},$$

$$z_h^{kj_k} \in \{0,1\}, h=1,2,\ldots,d,$$

$$\sum_{h=1}^d z_h^{kj_k} = 1, k=1,2,\ldots,i-1,$$

$$y_{i+1}(s^{-i}) \geq \frac{h-1}{d}y_{i-1}(s^{-i}) - (1-z_h^{i+1,j_{i+1}}),$$

$$y_{i+1}(s^{-i}) \geq \frac{h}{d}y_{i-1}(s^{-i}) + x_{j_{i+1}}^{i+1} - \frac{h}{d} - (1-z_h^{i+1,j_{i+1}}),$$

$$y_{i+1}(s^{-i}) \leq \frac{h-1}{d}y_{i-1}(s^{-i}) + x_{j_{i+1}}^{i+1} - \frac{h-1}{d} + (1-z_h^{i+1,j_{i+1}}),$$

$$y_{i+1}(s^{-i}) \leq \frac{h}{d}y_{i-1}(s^{-i}) + (1-z_h^{i+1,j_{i+1}}),$$

$$\frac{h-1}{d}z_h^{i+1,j_{i+1}} \leq x_{j_{i+1}}^{i+1} \leq z_h^{i+1,j_{i+1}}\frac{h}{d}+1-z_h^{i+1,j_{i+1}},$$

$$z_h^{i+1,j_{i+1}} \in \{0,1\},$$

$$h=1,2,\ldots,d,$$

$$\sum_{h=1}^d z_h^{i+1,j_{i+1}} = 1, \tag{3.7}$$

$$y_k(s^{-i}) \geq \frac{h-1}{d}y_{k-1}(s^{-i}) - (1-z_h^{kj_k}),$$

$$y_k(s^{-i}) \geq \frac{h}{d}y_{k-1}(s^{-i}) + x_{j_k}^k - \frac{h}{d} - (1-z_h^{kj_k}),$$

$$y_k(s^{-i}) \leq \frac{h-1}{d}y_{k-1}(s^{-i}) + x_{j_k}^k - \frac{h-1}{d} + (1-z_h^{kj_k}),$$

$$y_k(s^{-i}) \leq \frac{h}{d}y_{k-1}(s^{-i}) + (1-z_h^{kj_k}),$$

$$\frac{h-1}{d}z_h^{kj_k} \leq x_{j_k}^k \leq z_h^{kj_k}\frac{h}{d}+1-z_h^{kj_k},$$

$$z_h^{kj_k} \in \{0,1\}, h=1,2,\ldots,d,$$

$$\sum_{h=1}^d z_h^{kj_k} = 1, k=i+2,i+3,\ldots,n,$$

$$q(s^{-i}) = \begin{cases} y_n(s^{-i}) & \text{if } i<n, \\ y_{n-1}(s^{-i}) & \text{if } i=n, \end{cases}$$

$$s^{-i} \in S^{-i}, \ i=1,2,\ldots,n.$$

Therefore we can solve this mixed-integer linear programming system (3.7) to find all mixed-strategy Nash equilibria in normal form as described in system (3.1). The above idea is illustrated by the following example.

Example 4. Consider a three-player game $\Gamma = (N, S, \{u^i\}_{i\in N})$, where $N = \{1,2,3\}$, $S^i = \{s_1^i, s_2^i\}$, $i \in N$, and $\{u^i\}_{i\in N}$ are given by

	s_1^2	s_2^2	s_1^2	s_2^2
s_1^1	(1,1,1)	(1,0,1)	(0,1,0)	(1,0,0)
s_2^1	(1,1,1)	(1,0,1)	(0,1,0)	(0,0,0)
	s_1^3		s_2^3	

Let $\beta = 1$ and $d = 10$. The mixed-integer linear program for this game is given by

$$q(s_1^2, s_1^3) + q(s_2^2, s_1^3) + q(s_2^2, s_2^3) + \lambda_1^1 - \mu_1 = 0,$$
$$q(s_1^2, s_1^3) + q(s_2^2, s_1^3) + \lambda_2^1 - \mu_1 = 0,$$
$$x_1^1 + x_2^1 = 1,$$
$$q(s_1^1, s_1^3) + q(s_1^1, s_2^3) + q(s_2^1, s_1^3) + q(s_2^1, s_2^3) + \lambda_1^2 - \mu_2 = 0,$$
$$\lambda_2^2 - \mu_2 = 0,$$
$$x_1^2 + x_2^2 = 1,$$
$$q(s_1^1, s_1^2) + q(s_1^1, s_2^2) + q(s_2^1, s_1^2) + q(s_2^1, s_2^2) + \lambda_1^3 - \mu_3 = 0,$$
$$\lambda_2^3 - \mu_3 = 0,$$
$$x_1^3 + x_2^3 = 1,$$
$$x_j^i \leq \nu_j^i,$$
$$\lambda_j^i \leq 1 - \nu_j^i,$$
$$0 \leq x_j^i \leq 1,$$
$$\nu_j^i \in \{0, 1\},$$
$$\lambda_j^i \geq 0, \ j = 1, 2, \ i = 1, 2, 3,$$

$$
\begin{cases}
y_0(s_1^2, s_1^3) = 1, \\
y_2(s_1^2, s_1^3) \geq \frac{h-1}{10} y_0(s_1^2, s_1^3) - (1 - z_h^{21}), \\
y_2(s_1^2, s_1^3) \geq \frac{h}{10} y_0(s_1^2, s_1^3) + x_1^2 - \frac{h}{10} - (1 - z_h^{21}), \\
y_2(s_1^2, s_1^3) \leq \frac{h-1}{10} y_0(s_1^2, s_1^3) + x_1^2 - \frac{h-1}{10} + (1 - z_h^{21}), \\
y_2(s_1^2, s_1^3) \leq \frac{h}{10} y_0(s_1^2, s_1^3) + (1 - z_h^{21}), \\
\frac{h-1}{10} z_h^{21} \leq x_1^2 \leq z_h^{21} \frac{h}{10} + 1 - z_h^{21}, \\
h = 1, 2, \ldots, 10, \\
\sum_{h=1}^{10} z_h^{21} = 1, \\
y_3(s_1^2, s_1^3) \geq \frac{h-1}{10} y_2(s_1^2, s_1^3) - (1 - z_h^{31}), \\
y_3(s_1^2, s_1^3) \geq \frac{h}{10} y_2(s_1^2, s_1^3) + x_1^3 - \frac{h}{10} - (1 - z_h^{31}), \\
y_3(s_1^2, s_1^3) \leq \frac{h-1}{10} y_2(s_1^2, s_1^3) + x_1^3 - \frac{h-1}{10} + (1 - z_h^{31}), \\
y_3(s_1^2, s_1^3) \leq \frac{h}{10} y_2(s_1^2, s_1^3) + (1 - z_h^{31}), \\
\frac{h-1}{10} z_h^{31} \leq x_1^3 \leq z_h^{31} \frac{h}{10} + 1 - z_h^{31}, \\
h = 1, 2, \ldots, 10, \\
\sum_{h=1}^{10} z_h^{31} = 1, \\
q(s_1^2, s_1^3) = y_3(s_1^2, s_1^3).
\end{cases}
\tag{3.8}
$$

Similarly, $q(s_1^2, s_2^3)$, $q(s_2^2, s_1^3)$, $q(s_2^2, s_2^3)$, $q(s_1^1, s_1^3)$, $q(s_1^1, s_2^3)$, $q(s_2^1, s_1^3)$, $q(s_2^1, s_2^3)$, $q(s_1^1, s_1^2)$, $q(s_1^1, s_2^2)$, $q(s_2^1, s_1^2)$, and $q(s_2^1, s_2^2)$ can be obtained as $q(s_1^2, s_1^3)$ in system (3.8).

3.2 Numerical results

In this section, we present some numerical results. We use C++ to call ILOG CPLEX API functions to solve the mixed-integer linear programming (3.7). The MIP (mixed integer programming) search method, which is a dynamic search strategy or branch-and-cut strategy in ILOG CPLEX, is automatically determined by the ILOG CPLEX. All other parameters are also automatically set by ILOG CPLEX itself.

The code is run on a workstation of Lenovo ThinkStation D20 4155-BM4 with 16 processors and 16 G RAM. In the presentation of numerical results, we use the following symbols.

NumN: The number of players.

NumS: The number of strategies for each player.

NumEquilibra: The number of pure-strategy Nash equilibria for the instance.

Time: The total computational time to solve the problem.

Only a three-player game needs to be considered since any n-player game can be reduced to a three-player game in polynomial time as shown in [3]. In this chapter, we give the following randomly generated computation examples.

Example 5. Consider a three-player game $\Gamma = (N, S, \{u^i\}_{i \in N})$, where $N = \{1, 2, 3\}$. The number of strategies for each player, *NumS*, is randomly generated from 3 to 20. The $\{u^i\}_{i \in N}$ are randomly generated from 0 to 10.

Let $d = 10$ and $\beta = 1000$. The results can be given in Table 3.1. From the table we can see that this mixed-integer programming can successfully find all Nash equilibria in normal form. Although the largest dimension of these examples is only 3×19, this mixed-integer linear programming approach can solve the one of large dimension, and there are still many methods to speed up the computation process.

In these examples, only three-player games have been considered. When the number of player is greater than 3, we can use the polynomial-time method in [3] to convert this problem to a three-player game.

Table 3.1 The payoff function $u^i(s^i_j, s^{-i})$ randomly generated from 0 to 10.

Prob.	NumN	NumS	NumEquilibra	Time (s)
1	3	4	25	1.18
2	3	4	21	1.05
3	3	4	35	1.35
4	3	5	14	0.99
5	3	6	18	1.42
6	3	6	16	1.58
7	3	7	18	1.79
8	3	8	18	2.49
9	3	8	15	2.09
10	3	8	15	1.96
11	3	10	13	3.69
12	3	11	15	4.67
13	3	12	13	5.35
14	3	14	13	7.06
15	3	14	12	7.16
16	3	15	14	9.28
17	3	19	11	19.12
18	3	19	12	21.48

3.3 Summary

In this chapter, a mixed integer linear programming has been formulated to find all mixed-strategy Nash equilibria of a finite game in normal form, and some numerical results of this computing method have been given. More details about the mixed integer linear programming can be found in [4].

We first introduce an approximation of multilinear term in the first section. Based on the approximation and properties of mixed strategy, a mixed integer linear programming has been developed. By solving this mixed integer programming all Nash equilibria can be obtained. The example given in the first section explains the computing process well. Some numerical results of computing Nash equilibrium presented in the second section are promising.

References

[1] James Luedtke, Mahdi Namazifar, Jeff Linderoth, Some results on the strength of relaxations of multilinear functions, Mathematical Programming 136 (2) (2012) 325–351.

[2] Faiz A. Al-Khayyal, James E. Falk, Jointly constrained biconvex programming, Mathematics of Operations Research 8 (2) (1983) 273–286.

[3] V. Bubelis, On equilibria in finite games, International Journal of Game Theory 8 (2) (1979) 65–79.

[4] Zhengtian Wu, Chuangyin Dang, Fuyuan Hu, Baochuan Fu, A new method to finding all Nash equilibria, in: International Conference on Intelligent Science and Big Data Engineering, Springer, 2015, pp. 499–507.

CHAPTER 4

Solving long-haul airline disruption problem caused by groundings using a distributed fixed-point approach

4.1 Introduction

4.1.1 Briefly about airline disruption problem

Travel by aircrafts is the most convenient way of transportation, especially for long-distance travels. The operation of an airline requires the development of their flight schedules by creating flight scheduling, fleet assignment, aircraft routing, and crew scheduling, and the execution of airline operation is identical with the flight schedules [2]. However, frequently, there are disruptions that prevent these schedules from operating as originally planned. These disruptions are mainly caused by unanticipated incidents such as severe weather, crew absence, aircraft breakdowns, airport and air traffic restrictions, and so on. Disruptions of one flight or airport may spread to the following flights and airports causing massive flight cancellations and delays [3], and the European airline punctuality report [4] reveals that only 3.6% of flights are delayed by severe weather condition, whereas 15.1% of fights are delayed due to the propagation of the delayed flights [3]. That is to say, the solution methods and efficiency of the airline disruption problem affect the whole planned schedule more than sources of disruptions. So an airline must be able to return to the original schedule from disruptions by producing the recovery plan as quickly as possible.

The recovery plan is produced by reassigning crews, aircrafts, passengers, and other related resources, and as a result, some flights are canceled, and some are delays according to the recovery plan. The recovery plan is required to be less deviated from the original plan, and it can be recovered from disruption to original schedule quickly. An inefficient recovery plan results in more aircrafts being delayed or canceled. It is a complex task to produce recovery plans because many resources such as passengers, aircraft, and crew have to be reassigned. When a disruption occurs in the day of

Integer Optimization and Its Computation in Emergency Management
https://doi.org/10.1016/B978-0-32-395203-3.00009-5

operation, large airlines always respond by calculating the problem in a sequential fashion concerning the problem components: passengers, crews, ground operations, and aircrafts. The whole process is iterated until a new schedule for all the aircrafts, crews, ground operations, and passengers can be implemented [2], and the original schedule can be resumed when the disruption is over.

4.1.2 Literature review

Teodorović and Guberinić [5] are pioneers in researching this airline disruption problem. They aim at minimizing the total passenger delay by branch-and-bound methods. Later, Teodorović and Stojković [6] give a lexicographic dynamic programming scheme by minimizing the number of the total passenger delays and cancellations. Teodorović and Stojković [7] propose a more complete solution process to solve this problem by including all operational constraints. None of the above papers considers flight delay costs and cancellation costs. Jarrah et al. [8] propose two models based on minimum cost network flow: one minimizes flight delays, and the other minimizes flight cancellations. However, they cannot treat delays and cancellations together in one model. Yu [9] also provides a network model of the problem, which appears to be NP-hard, but gives no solution method. His ideas are further developed by Argüello [10] and Argüello et al. [11,12], who present a time-band optimization models and a greedy randomized adaptive search procedure to solve that problem by minimizing the total cost for delays and cancellations. Lettovský [13] presents a framework for integrated recovery that considers aircraft, crew, and passenger together. A matter problem is proposed to control the previous three subproblems, and only parts of implementations are provided in his paper. Thengvall et al. [14] propose a perturbation model considering aircraft shortages, which can treat cancellations, delays and flight swaps together without considering crew or maintenance. Bard et al. [15] develop the time-band optimization model based on [10] for rescheduling aircraft routes in response to delays and groundings in the disruption. Løve et al. [16] implement a local search heuristic method capable of generating feasible revised flight schedules of high quality in less than 10 seconds. The instances used in the paper are created by a generator, not from practical airline schedule. Rosenberger et al. [17] present an optimization model that reschedules flight routes by minimizing an objective function about cancellation costs. The model contains a route generation procedure and a set-partitioning problem. Andersson and Värbrand [18] propose a mixed integer multicommodity flow model with side constraints, which is improved to be a set-packing model by the Dantzig–Wolfe decomposition. Two instants are tested, but the larger instant needs much more time than the smaller instant. Yan et al. [19] develop a two-step framework to construct a real-time schedule. Kohl et al. [20] give an introduction of airline disruption management and experiences from project Descartes. Jeng [3] develops an multiobjective inequality-based genetic algorithm (MMGA), which combines the genetic algorithm and is able to solve multiple-objective airline disruption problems. For more theoretical descriptions and comparisons of the academic research on airline disruption management, we refer to the review by Clausen et al. [2], PhD thesis by Jeng [3], and books by Wu [21], Bazargan [22], Teodorović [23], Abdelghany et al. [24], and Yu [25].

In this chapter, to use distributed computation, the airline disruption problem is modeled as two subproblems: feasible flight routes generation in Subsection 4.2.1 and aircrafts reassignment in Subsection 4.2.2. A distributed computation is proposed based on Dang's algorithm [1] for integer programming to produce feasible flight routes by dividing the solution space into several segments using two division methods. In this chapter, we mainly focus on the second division method. Beginning with the original flight routes, the generation for new feasible flight routes starts from two directions that generate feasible flight routes, both larger than or equal to and less than or equal to the original flight routes in lexicographical order. In lexicographical order, these later generated feasible flight routes increase or decrease from earlier generated feasible flight routes, so feasible flight routes, which are obtained earlier, are deviated less from the original flight routes than those later obtained. For a long-haul airline disruption problem, in a reasonable time, it is very hard to generate all feasible flight routes. So only partial feasible flight routes are produced in each segment. Comparisons of these partial feasible flight routes generated by Dang's algorithm and CPLEX CP Optimizer show that partial feasible flight routes generated by the former method outperform those generated by the latter one when these partial feasible flight routes are used to solve long-haul aircrafts reassignment problem by CPLEX Optimizers's Concert Technology. As far as we know from the literature, this distributed implementation is the first one for feasible flight routes generation.

The remaining part of this chapter is organized as follows. Section 4.2 presents the mathematical formulation of the two subproblems. Two division methods are proposed to divide the solution space and conduct a distributed computation in Section 4.3. In Section 4.4, numerical results

Table 4.1 Sample schedule.

Aircraft	Flight ID	Origin	Destination	Departure time	Arrival time	Duration	Cancellation cost
A	11	BOI	SEA	1410	1520	1:10	7350
	12	SEA	GEG	1605	1700	0:55	10 231
	13	GEG	SEA	1740	1840	1:00	7434
	14	SEA	BOI	1920	2035	1:15	14 191
B	21	SEA	BOI	1545	1700	1:15	11 189
	22	BOI	SEA	1740	1850	1:10	12 985
	23	SEA	GEG	1930	2030	1:00	11 491
	24	GEG	SEA	2115	2215	1:00	9581
C	31	GEG	PDX	1515	1620	1:05	9996
	32	PDX	GEG	1730	1830	1:00	15 180
	33	GEG	PDX	1910	2020	1:10	17 375
	34	PDX	GEG	2100	2155	0:55	15 624

of comparing performance among these partial feasible flight routes generated by Dang's algorithm and CPLEX CP Optimizer when solving aircraft reassignment problem are presented. Section 4.5 offers conclusions of this chapter.

4.2 Problem formulation

Considering a situation that a disruption happens in the execution of the original schedule, resulting one or more aircrafts are grounded and unable to fly during the whole day. All flights are reassigned to remaining available aircrafts, and the original schedule is required to be resumable at the end of the disruption period. With the objective of finding an aircraft reassignment that minimizes the lost of disruption and deviation from the original schedule, the process to solve this disruption problem concerns two subproblems. Feasible flight routes for all available aircrafts are first generated in the first subproblem defined in Subsection 4.2.1, and then all these feasible flight routes are reassigned to remaining available aircrafts in the second subproblem in Subsection 4.2.2. A sample schedule [26] introduced in Table 4.1 is used to illustrate these two subproblems.

The sample schedule presented here assigns 3 aircrafts to service 12 flights over 4 stations. Considering a disruption occurs due to the grounding of aircraft B, thus flights from 21 to 24 will be canceled if no reschedule is made, and that will result in a total cost about 45 246. If all these flights are reassigned to remaining aircrafts A and C, then a new schedule can be obtained of which the total cost is less than or equal to the old one. Assuming a departure curfew time and no arrival curfew time [26], we further present the details of these two subproblems to solve the airline disruption problem.

4.2.1 Feasible flight routes generation

In Fig. 4.1, we introduce a feasibility problem to produce feasible flight routes based on the connection network. The four stations are represented by four nodes, which is marked by letters. Flight routes between any two stations are marked by arcs. Two additional nodes, arcs linking them and station nodes, are added to present the origin and destination of each feasible flight route. There is no arc connecting source node and station node P or sink node and station node P, because no route starts from or terminates at node P. In this connection network, the number of arcs that flow to a station node equilibrates the number of arcs that flow out from the station node. The relations of sequences about time between any two flight legs are not revealed in this network. So a flight leg position in a feasible flight route cannot be decided so far. To solve all possible combinations of arcs shown in Fig. 4.1 under several constraints, we give the model below. This model is relatively simpler and easy to solve because sequences of flights are not decided. A heuristics transformation is introduced later to transform the solutions of this model into actual flight routes, and the right sequences of these flight legs are decided through the transformation.

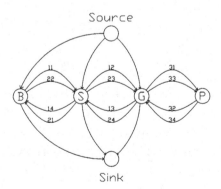

Figure 4.1 Connection network representation of the problem.

In the formulation, we use the following notations.

Indices

 i, j, k flight indices

 t station index

Sets

 F set of flights

 S set of stations

Parameters

 td_i duration of flight i

 tt_i turnaround time of flight i

 T_j total time from departure of flight j to departure curfew time of station

 d_{it} is 1 if flight i departs from station t; 0 if not

 a_{it} is 1 if flight i arrives at station t; 0 if not

 est_{ijk} is 1 if flight k's destination and flight j's original are the same, and flight k's original and flight j's destination are the same; is -1 if flight i's destination and flight k's original are the same, or flight i's destination and flight k's destination are the same, or flight i's original and flight k's original are the same, or flight i's original and flight k's destination are the same; 0 if not

 en_{jt} is -1 if flight j's destination and station t are the same, or flight j's original and station t are the same; 0 if not

Variables

 x_i is 1 if flight leg i is contained in a feasible flight route; 0 if not

 s_t is 1 if the origin station in the feasible flight route is the station t; 0 if not

 k_t is 1 if the destination station in the feasible flight route is the station t; 0 if not

Mathematical formulation of feasible route generation

$$\exists x_i, s_t, k_t \in \{0, 1\} \quad \forall i \in F, \forall t \in S \qquad (4.1a)$$

subject to

(limitation on flight time) $\displaystyle\sum_{i=j}^{Card(F)-1} x_i(td_i + tt_i) \leq T_j \quad \forall j \in F, \qquad (4.1b)$

(node conservation) $\displaystyle\sum_{i\in F} x_i a_{it} - \sum_{i\in F} x_i d_{it} + s_t - k_t = 0 \quad \forall t \in S, \quad (4.1c)$

(flow from source node) $\displaystyle\sum_{i\in F} x_i d_{it} \geq s_t \quad \forall t \in S, \qquad (4.1d)$

(flow to sink node) $\qquad \sum_{i\in F} x_i a_{it} \geq k_t \quad \forall t \in S,$ $\qquad$ (4.1e)

(source node cover) $\qquad \sum_{t\in S} s_t = 1,$ $\qquad$ (4.1f)

(sink node cover) $\qquad \sum_{t\in S} k_t = 1,$ $\qquad$ (4.1g)

(subroute elimination) $\qquad \sum_{i\in F} x_i est_{ijk} + \sum_{t\in S} s_t en_{jt} + \sum_{t\in S} k_t en_{jt} \geq 1 \quad \forall j \in F,$

$$\forall k \in (j+1, \ldots, Card(F) - 1). \qquad (4.1h)$$

Prior to the construction of this model, original flight legs are rearranged according to the departure time of each flight. So x_i, which indicate flight legs, are in the ascending order of departure time. The parameters td_i and tt_i and sets F and S are inputs, which are traced to the flight schedule. The parameters T_j can be calculated by subtracting flight i's departure time from the departure curfew time of stations. After reviewing each flight leg's origin and destination concerning each station, we can obtain the parameters d_{it}, a_{it}, est_{ijk}, and en_{jt}.

The purpose of this model is enumerating all feasible flight routes under conditions (4.1b)–(4.1g). Constraint (4.1b) ensures that the departure time of the last flight leg in each flight route does not exceed the departure curfew time of stations assuming that the x_i with the smallest subscript i is the first flight leg in a flight route for each j. Constraint (4.1c) ensures that the flow number coming into a station node has to equilibrate the flow number coming out from the same one. Constraints (4.1d) and (4.1e) are formulated to ensure that if there is a flow that comes from the source node to a station one, then there has to be at least one flow that comes from the previous station node to another one, or if there is a flow that comes from a station node to the sink one, then there has to be at least one flow that comes from another station node to the previous one. Constraints (4.1f) and (4.1g) are developed to prompt that there is only one flow that comes from the source node to a station one and only one flow that comes from a station node to the sink one. Constraint (4.1h) eliminates subroutes (circles of flight legs) containing only two flight legs. Constraints eliminating subroutes containing more than two flight legs can be easily constructed like constraint (4.1h), but that results in a huge increase in the total amount of constraints that are inefficient for computation. So constraints eliminating subroutes containing more than two flight legs are not considered here, and these subroutes containing more than two flight legs can be eliminated by the transformation in Fig. 4.2.

The set of all feasible flight routes is marked by P. It is stated in [10] that the cardinality of P ($Card(P)$) is much larger concerning to $Card(F)$ and $Card(S)$. If the length of each flight route is limited to be not more than v flight legs, where $v \ll Card(F)$ and $v < Card(S)$, then it is stated in [10] that $Card(P)$ is bounded below by the function

$$\Phi(v) = O(2^v). \tag{4.2}$$

Therefore $Card(P)$ is exponential concerning to the largest flight route length v. No detail of generating P is presented in [10], [11], and [17], so a feasibility problem (4.1) is introduced here to produce P. The CPLEX CP Optimizer can solve all feasible flight routes easily for small $Card(F)$ and $Card(S)$. However, when $Card(F)$ and $Card(S)$ are large, $Card(P)$ is extremely large, and CPLEX CP Optimizer is not able to solve all feasible flight routes in a short time. Based on the integer programming developed in [1], a distributed computational method is introduced to solve feasible flight routes. The result of the distributed method outperforms those produced by CPLEX CP Optimizer. Some details of this approach are presented in Section 4.3.

The solutions generated from problem (4.1) are simply binary numbers. A heuristics procedure is applied to transform these binary numbers into actual feasible flight routes. Containers are first defined to save these actual feasible flight routes. Feasible solution values for variables s_t and k_t are inspected to decide the origin and destination in each flight route. The chosen flight legs are selected out from x_i in these binary solutions. The chosen flight legs whose origin is the same as the origin of this flight route are saved in the back of a flight route container. So the amount of feasible flight route containers is equal to the amount of flight legs whose origin is equal to the origin of this flight route. For each flight route container, the chosen flight legs that are not in the flight route container are saved in the back of the flight route container if the origin of the new added flight leg is the same as the destination of the last flight leg in the flight route container. If there is more than one flight leg whose origin is the same as the destination of the last flight leg in the flight route container, then the flight route container is duplicated into the amount equal to the amount of these to be added flight legs. This process is iterated till there is no flight leg that can be saved in the back of a flight route container. The flight route

k_{tj} is 1 if the destination station of the feasible flight route j is station t; 0 if not

dc_j the feasible flight route j's delay cost

cc_i the flight i's cancellation cost

p_j the feasible flight route j's profit

h_t number of aircrafts that have to departure from source stations t

g_t number of aircrafts that have to terminate at sink stations t

TN total number of aircrafts

GN grounding number of aircrafts

Variables

y_j is 1 if feasible flight route j is rescheduled to an aircraft; 0 if not

z_i is 1 if flight i is canceled; 0 if not

Mathematical formulation of aircrafts reassignment

minimize
$$\sum_{j \in P} dc_j y_j + \sum_{i \in F} cc_i z_i \qquad (4.3\text{a})$$

or maximize (alternative)
$$\sum_{j \in P} p_j y_j \qquad (4.3\text{b})$$

subject to

(flight cover)
$$\sum_{j \in P} x_{ij} y_j + z_i = 1 \qquad \forall i \in F, \qquad (4.3\text{c})$$

(aircrafts balance of source stations)
$$\sum_{j \in P} s_{tj} y_j = h_t \qquad \forall t \in S, \qquad (4.3\text{d})$$

(aircrafts balance of sink stations)
$$\sum_{j \in P} k_{tj} y_j = g_t \qquad \forall t \in S, \qquad (4.3\text{e})$$

(total available aircrafts)
$$\sum_{j \in P} y_j = TN - GN, \qquad (4.3\text{f})$$

(binary assignment)
$$y_j \in 0, 1 \qquad \forall j \in P, \qquad (4.3\text{g})$$

(binary assignment)
$$z_i \in 0, 1 \qquad \forall i \in F. \qquad (4.3\text{h})$$

The parameters x_{ij}, s_{tj}, and k_{tj} are feasible solutions of Subsection 4.2.1. By p_j we denote the sums of revenues of each individual flight leg that constitutes flight routes minus the delay cost of each flight route plus the bonus per flight leg covered. The remaining parameters are input parameters that date from the flight schedule.

The objective is represented by (4.3a). On one hand, this objective minimizes the sum delay cost of every chosen flight route. On the other hand, it minimizes the sum cancellation cost of every canceled flight leg. The objective (4.3b) maximizes the total profit of each chosen flight route. Because there is a bonus in p_j that encourages the flight route to deviate less from the original flight route for the long-haul problem in Subsection 4.4.2, the results for minimizing the cost and maximizing the profit are not the same. So both objective functions participate in computation of the long-haul problem in Subsection 4.4.2. No bonus in p_j is given in the small problem in Subsection 4.4.1, so only objective (4.3a) is tested for the small problem in Subsection 4.4.1. Flight cover constraint (4.3c) ensures that all flight legs have to be either canceled or in a flight route. Constraints (4.3d) and (4.3e), which are called aircrafts balance of source stations constraint and sink stations constraint, respectively, ensure the requirements for the numbers of aircrafts located at the beginning and at the end of disruption period for every station. Constraint (4.3f) implements the requirement that the total number of chosen feasible flight routes is the same as the number of available aircrafts. Fractional solutions are precluded by constraints (4.3g) and (4.3h).

Problem (4.3) is solved by CPLEX Optimizers's Concert Technology, and these solutions are combinations of feasible flight routes. Sometimes, there exist more than one optimum solution, but the amount of original flight routes differs among these optima. A combination of feasible flight routes that includes more original flight routes hints that the new schedule is less different from the original schedule. A simple integer programming is developed here to find out the combination with the most original flight routes.

We will use the following notations in the formulation.

Indices

    i feasible flight routes' combination indices

Sets

    C set of feasible flight routes' combinations

Parameters

    AOR_i amount of original flight route in feasible flight routes' combination i

Variables

u_i 1 if feasible flight routes' combination i contains maximum original flight route; 0 otherwise

Least deviation mathematical formulation

$$\text{maximize} \quad \sum_{i \in C} AOR_i u_i \tag{4.4a}$$

subject to

$$\sum_{i \in C} u_i = 1, \tag{4.4b}$$

$$u_i \in 0, 1 \quad \forall i \in C. \tag{4.4c}$$

The parameter AOR_i can be easily determined by inspecting how many original flight routes each feasible flight routes' combination has.

4.3 Methodology

Both the formulation of problem (4.1) in Subsection 4.2.1 and the formulation of problem (4.3) in Subsection 4.2.2 can be solved by CPLEX, but it is difficult to find all feasible flight routes for problem (4.1) in a short time according to Subsection 4.2.1. There are 162 flights in the data set in Subsection 4.4.2. Therefore, according to Eq. (4.2), the upper bound of the number of feasible flight routes is 2^{162} in problem (4.1). The upper time bound of computing all feasible flight routes of problem (4.1) for the data set in Subsection 4.4.2 using CPLEX CP Optimizer is obtained through a simple calculation. It costs 21762 ms to find 100 000 feasible flight routes by CPLEX CP Optimizer. The average time to obtain a feasible flight route is at least 0.21762 ms because it costs more time to find a feasible flight route as the search tree goes deeper. For computing all these 2^{162} feasible flight routes, it costs at least 1.27×10^{45} s, which is about 4×10^{37} years in the worst case by CPLEX CP Optimizer.

The distributed implementation of Dang's algorithm is developed to produce feasible flight routes after the solution space of problem (4.1) is divided into some segments. It works well especially when the solution space is very large. A simple example is first used to illustrate Dang's algorithm. Given a polytope $P = \{x \in \mathbb{R}^2 \| Ax \leq b\}$ with

$$A = \begin{pmatrix} -17 & 2 \\ 6 & 5 \\ -3 & -3 \end{pmatrix} \text{ and } b = \begin{pmatrix} -8 \\ 4 \\ 7 \end{pmatrix}$$

and a lattice $D(P) = \{x \in \mathbb{Z}^2 | x^l \leq x \leq x^u\}$ as illustrated in Fig. 4.3, the problem is to find an $x \in P \cap D(P)$ or prove that there is no such point. The idea to solve this problem is to define an increasing mapping from the lattice into itself. The integer points outside P are mapped into the first point in P that is smaller than they in lexicographical order or x^l. Under this increasing mapping, all integer points that inside the polytope are fixed points. Given an initial integer point, the approach is either finding an integer point in the polytope or proving that no such point exists. With a simple modification, all integer points in P can be obtained sequentially. More details on this iterative approach can be found in [1].

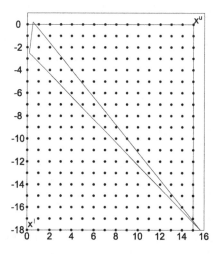

Figure 4.3 Polytope P and lattice.

The distributed implementation of Dang's algorithm consists of two parts. The first part is to divide the solution space into some segments. The other part is to find integer points in each segment by Dang's algorithm. Each segment obtained in the first part is independent of each other. Therefore in each segment the calculation of integer points can proceed at the same time in each processor. Two division methods are provided here to solve the airline disruption problem. We attempt to divide the solution space into several segments that contain more or less equal integer points in $D(P)$. For a long-haul problem, sufficient processors are needed to divide the solution space into segments that can finish computing in a reasonable time. If the computation of each segment still lasts an extremely long time, and no more processors can be provided, then we propose another division method. This division method is specially devised to solve the airline disruption problem using original flight routes as initial points. Two bound points, which are in the middle of two continuous initial points in lexico-graphical order for this division approach, define the segment. A cluster is defined around each initial point in each segment for computation, and a parameter is used to control the number of feasible flight routes solved in each cluster. Only a small percentage of all feasible flight routes of prob-lem (4.1) produced by Dang's algorithm is necessary to solve problem (4.3) better than canceling the flight route original assigned to the grounding aircraft. Details of these two division methods are presented below.

4.3.1 Nearly average division

All integer points in $D(P)$ in Fig. 4.3 can be reordered as an one-dimensional chain based on lexicographic order. Thus no matter what the dimension of the problem space is, it can be treated as one-dimensional. Since all integer points are in a chain, the chain can be divided into more or less equal segments.

Let N_s be the amount of segments that $D(P)$ is divided into. The first $Card(D(P))$ mod N_s segments are allocated $\lfloor \frac{Card(D(P))}{N_s} + 1 \rfloor$ integer points in $D(P)$ to compute, and the remaining $N_s - Card(D(P))$ mod N_s segments are allocated $\lfloor \frac{Card(D(P))}{N_s} \rfloor$ integer points in $D(P)$ to compute. With this division and allocation, the maximum difference of the amount of integer points between any segment is just one. So all points in $D(P)$ are divided into several segments nearly averagely. The sum of the amount of integer points in each segment is exactly equal to the total integer points in $D(P)$ through this allocation. A simple deduction is applied below to testify the equality

$$(Card(D(P)) \bmod N_s) \times \lfloor \frac{Card(D(P))}{N_s} + 1 \rfloor + (N_s - Card(D(P)) \bmod N_s)$$
$$\times \lfloor \frac{Card(D(P))}{N_s} \rfloor$$
$$= N_s \times \lfloor \frac{Card(D(P))}{N_s} \rfloor + (Card(D(P)) \bmod N_s)$$
$$= Card(D(P)).$$

For instance, $D(P)$ is divided into five segments distinguished by dots and crosses as illustrated in Fig. 4.4. Four segments contain 61 integer points, and one segment contains 60 integer points where total amount of integer points is 304.

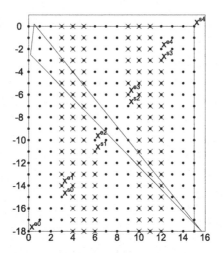

Figure 4.4 Near average division.

Let x^{e_i} and x^{s_i} define each segment as noted in Fig. 4.4, and let $D(P) = \bigcup \{x \in D(P) | x^{e_i} \leq_l x \leq_l x^{s_i}\}$. Starting from each x^{s_i}, all feasible solutions in each $P \bigcap \{x \in D(P) | x^{e_i} \leq_l x \leq_l x^{s_i}\}$ can be obtained simultaneously by this distributed computation. Since it is easy to obtain $Card(\{x \in D(P) | x^{e_i} \leq_l x \leq_l x^{s_i}\})$, the importance is to calculate values of x^{e_i} and x^{s_i} that define each segment.

Lemma 5. Let $x^u = (x_0^u, x_1^u, \ldots, x_i^u, \ldots, x_{n-1}^u)$, $x^l = (x_0^l, x_1^l, \ldots, x_i^l, \ldots, x_{n-1}^l)$ $\leq_l x^u$, $p_i = x_i^u - x_i^l + 1$ for $i = 0, 1, \ldots, n-1$, and $s = Card(D_u)$ or $s = Card(D_l)$, where $D_u = \{x \in D(p) \mid x^s \leq_l x <_l x^u\}$ and $D_l = \{x \in D(p) \mid x^l <_l x \leq_l x^s\}$.

Then any integer point $x^s = (x_0^s, x_1^s, \cdots, x_i^s, \cdots, x_{n-1}^s)$ in $D(P)$ can be calculated by the following formula:

$$
x_i^s = \begin{cases} x_i^u - \left\lfloor \dfrac{s \bmod \prod\limits_{j=i}^{n-1} p_j}{\prod\limits_{j=i+1}^{n-1} p_j} \right\rfloor & \text{for } i = 0, 1, \ldots, n-2, \\ x_{n-1}^u - s \bmod p_{n-1} & \text{for } i = n-1, \end{cases} \tag{4.5}
$$

or

$$
x_i^s = \begin{cases} x_i^l + \left\lfloor \dfrac{s \bmod \prod\limits_{j=i}^{n-1} p_j}{\prod\limits_{j=i+1}^{n-1} p_j} \right\rfloor & \text{for } i = 0, 1, \ldots, n-2, \\ x_{n-1}^l + s \bmod p_{n-1} & \text{for } i = n-1. \end{cases} \tag{4.6}
$$

Proof. Given any integer point x^s in $D(P)$ and x^u, let

$$
D_i^u = \begin{cases}
\{x \in D(p) \mid x_i^s + 1 \le x_i \le x_i^u\} \\
\quad \text{for } i = 0, \\
\{x \in D(p) \mid x_i^s + 1 \le x_i \le x_i^u, \text{ and } x_{j-1} = x_{j-1}^s \text{ for } j = 1, 2, \ldots, i\} \\
\quad \text{for } i = 1, 2, \ldots, n-1.
\end{cases}
$$

Obviously,

$$
D_u = \bigcup_{i=0}^{n-1} D_i^u. \tag{4.7}
$$

So

$$
\begin{aligned}
s = Card(D_u) &= \sum_{i=0}^{n-1} Card(D_i^u) \\
&= \sum_{i=0}^{n-2} \left((x_i^u - x_i^s) \prod_{j=i+1}^{n-1} p_j \right) + (x_{n-1}^u - x_{n-1}^s) \\
&= (x_0^u - x_0^s) \prod_{j=1}^{n-1} p_j + (x_1^u - x_1^s) \prod_{j=2}^{n-1} p_j + \cdots \\
&\quad + (x_{n-2}^u - x_{n-2}^s) \prod_{j=n-1}^{n-1} p_j + (x_{n-1}^u - x_{n-1}^s).
\end{aligned} \tag{4.8}
$$

Since $p_i = x_i^u - x_i^l + 1$, we have $(x_i^u - x_i^s) < p_i$ for $i = 0, 1, \ldots, n-1$. Then

$$
(x_i^u - x_i^s) \prod_{j=i+1}^{n-1} p_j < p_i \prod_{j=i+1}^{n-1} p_j = \prod_{j=i}^{n-1} p_j, \tag{4.9}
$$

$$
(x_{i+1}^u - x_{i+1}^s) \prod_{j=i+2}^{n-1} p_j < p_{i+1} \prod_{j=i+2}^{n-1} p_j = \prod_{j=i+1}^{n-1} p_j \le \prod_{j=i}^{n-1} p_j, \tag{4.10}
$$

$$
(x_{i+2}^u - x_{i+2}^s) \prod_{j=i+3}^{n-1} p_j < p_{i+2} \prod_{j=i+3}^{n-1} p_j = \prod_{j=i+2}^{n-1} p_j \le \prod_{j=i+1}^{n-1} p_j \le \prod_{j=i}^{n-1} p_j, \tag{4.11}
$$

$$
\cdots \cdots \cdots
$$

$$
(x_{n-2}^u - x_{n-2}^s) \prod_{j=n-1}^{n-1} p_j < p_{n-2} \prod_{j=n-1}^{n-1} p_j = \prod_{j=n-2}^{n-1} p_j \le \prod_{j=i+1}^{n-1} p_j \le \prod_{j=i}^{n-1} p_j. \tag{4.12}
$$

The following result can be deducted by using (4.9)–(4.12):

$$
\begin{cases}
s \bmod \prod_{j=i}^{n-1} p_j = \sum_{l=i}^{n-2}\left((x_i^u - x_i^s)\prod_{j=l+1}^{n-1} p_j\right) & \text{for } i = 0, 1, \ldots, n-2, \\
s \bmod p_{n-1} = x_{n-1}^u - x_{n-1}^s & \text{for } i = n-1.
\end{cases}
\tag{4.13}
$$

Finally, we obtain the result by using (4.9) to (4.12) again:

$$
\begin{cases}
\left\lfloor \dfrac{s \bmod \prod_{j=i}^{n-1} p_j}{\prod_{j=i+1}^{n-1} p_j} \right\rfloor = \left\lfloor \dfrac{\sum_{l=i}^{n-2}\left((x_i^u - x_i^s)\prod_{j=l+1}^{n-1} p_j\right)}{\prod_{j=i+1}^{n-1} p_j} \right\rfloor = x_i^u - x_i^s \\
\qquad \text{for } i = 0, 1, \ldots, n-2, \\
s \bmod p_{n-1} = x_{n-1}^u - x_{n-1}^s \quad \text{for } i = n-1.
\end{cases}
\tag{4.14}
$$

This completes the proof of (4.5). The proof of (4.6) is similar to that of (4.5) when x^s and x^l are given, so it is omitted here. With these x^{ei} and x^{si} calculated by (4.5) or (4.6), all feasible solutions in each segment defined by x^{ei} and x^{si} can be obtained simultaneously using a distributed computation network, the details of which are given in Subsection 4.3.3.

4.3.2 Initial seeds cluster division

For a long-haul airline disruption problem as in Subsection 4.4.2, in a short time, it is difficult to find all feasible flight routes even providing computation equipments with excellent hardware and conducting a distribution computation based on the division method in Subsection 4.3.1.

An alternative division method is developed to conquer the airline disruption problem with larger dimension. Because only a small percentage of all feasible flight routes of problem (4.1) produced by Dang's algorithm is combined with this division method, a solution of problem (4.3) that is better than canceling the flight route original assigned to the grounding aircraft can be obtained.

To produce new feasible flight routes, let the seeds be the original flight routes. Two bound points in the middle of two continuous initial points in lexicographical order define the segments. A cluster, which includes feasible flight routes, is defined around each seed in a segment. The parameter of cluster size is used to control the number of feasible flight routes to be received in each cluster as in Fig. 4.5. In lexicographic order, the feasible solutions obtained in Dang's algorithm are less than or equal to the initial point all the time. After a simple transformation of Dang's algorithm, feasible solutions that are larger than or equal to initial point in lexicographic order can be found. Therefore the feasible flight routes can be obtained in two directions beginning from the seed. As we can see in Fig. 4.5, once it is encountered by a bound point or the number of feasible flight routes produced is equal to the cluster size, the process of producing stops. After these bound points used to define these segments are given, the process of producing of feasible flight routes begins at the seed in each segment using Dang's algorithm at the same time in a distributed network presented in Subsection 4.3.3. CPLEX CP Optimizer can also be used to generate feasible flight routes in each segment defined by bound points. Comparisons of feasible flight routes generated by these two methods are given in Section 4.4.

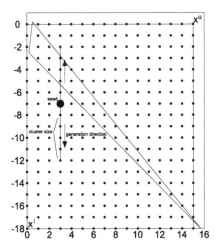

Figure 4.5 Initial seeds cluster division.

We present the computation process of these bound points. Let $x^{l_{i-1}}$, x^{l_i}, $x^{l_{i+1}}$ ($x^l <_l x^{l_{i-1}} <_l x^{l_i} <_l x^{l_{i+1}} <_l x^u$) be three continuous initial points seen as seeds, and thus there is no seed between $x^{l_{i-1}}$ and x^{l_i} or x^{l_i} and $x^{l_{i+1}}$. Let $x^{b_{i-1}}$ be the bound point in the middle of $x^{l_{i-1}}$ and x^{l_i} in lexicographical order, and let x^{b_i} be the bound point in the middle of x^{l_i} and $x^{l_{i+1}}$, and thus $x^l <_l x^{l_{i-1}} <_l x^{b_{i-1}} <_l x^{l_i} <_l x^{b_i} <_l x^{l_{i+1}} <_l x^u$. So x^{b_i} can be calculated by (4.5) given s^{b_i}, which is calculated as

$$
\begin{aligned}
s^{b_i} &= Card(\{x \in D(P)|x^{b_i} \leq_l x <_l x^u\}) \\
&= \left\lfloor \frac{Card(\{x \in D(P)|x^{l_{i+1}} \leq_l x <_l x^u\}) + Card(\{x \in D(P)|x^{l_i} \leq_l x <_l x^u\})}{2} \right\rfloor
\end{aligned}
$$

$$(4.15)$$

Each cluster is defined in the segment $\{x \in D(P)|x^{b_{i-1}} <_l x \leq_l x^{b_i}\}$, where $D(P) = \bigcup \{x \in D(P)|x^{b_{i-1}} <_l x \leq_l x^{b_i}\}$ with $x^{b-1} = x^l$, $x^{b_{Card(...,x^{l_i-1},x^{l_i},x^{l_i+1},...)-1}} = x^u$. Beginning at x^{l_i}, feasible flight routes whose amount is equal to the cluster size could be solved by Dang's algorithm along with increasing and decreasing direction in lexicographical order within the bound $(x^{b_{i-1}}, x^{b_i}]$. The first feasible flight route found in both directions from the seed is the least deviated from the seed flight route, and there is no other feasible flight route between two continuous feasible flight routes produced. The earlier a feasible flight route is obtained, the less this feasible flight route is deviated from x^{l_i}. Meanwhile, by the computation experience in Section 4.4 the new reassignment of aircrafts when disruption happens always contains feasible flight routes that are less deviated from the original route. These feasible flight routes, which are produced near original flight routes, not only satisfy the lower deviation of the reassignment from the original assignment but also can constitute a reassignment with a total cost less than the cost of canceling the original flight route assigned to the grounding aircraft. The computation for feasible flight routes in each cluster can also be done by CPLEX CP Optimizer through adding two lexicographic constraints to the formulation of problem (4.1) and restricting these generated feasible flight routes in the segment $\{x \in D(P)|x^{b_{i-1}} <_l x \leq_l x^{b_i}\}$, which contains the cluster that is being computed. There is no relation between the feasible flight routes produced by CPLEX CP Optimizer and the original flight routes. The sequence of these feasible flight routes obtained is orderless in lexicographic order and depends on the CPLEX CP Optimizer's search strategy. To find a reasonable scheduler for aircrafts, a mass of feasible flight routes is required to be produced by CPLEX CP Optimizer. However, only several outstanding feasible flight routes are required to be produced by Dang's algorithm from the computation experience in Section 4.4.

Each cluster $\{x \in D(P)|x^{b_{i-1}} <_l x \leq_l x^{b_i}\}$ can be divided again by the division algorithm introduced in Subsection 4.3.1 if more computation processors can be provided.

4.3.3 The implementation of distributed computation

It is independent among every resulting segment produced by the division method presented in Subsection 4.3.1 or Subsection 4.3.2. Therefore, for each segment, it can be computed simultaneously from the start point to the end one. In this implementation of distributed computation, the message passing interface (MPI) is used to build communication network among processors according to respective computers, and OpenMP is used to implement a parallel calculation among all processors in the same computer. The pseudocode of implementing the distributed computation network is described in Fig. 4.6. "MPI_Comm_rank", a function of MPI, indexes the computer's rank. It begins at 0 and is kept in a variable *computerid*. All computer's names are contained in the configuration file. In this configuration file, the master computer is ranked first because it has to take the charge of the procedure of distributed computation. Therefore the *computerid* 0 stands for the master computer. Slave computers are marked from 1 to the number of total slave computers. The master computer and each slave computer must have a copy of both configuration file and distributed computation program execution file. The MPI can execute all execution files at the same time in each computer. The variable *SystemInfo.dwNumberOfProcessors* is the sum of processors in the master computer and each slave computer. The solution space is divided by one of the two division methods in Section 4.3. Each slave computer receives each segment defined by the start point x^{s_i} and the end point x^{e_i} computed in the master computer. If a slave computer has more processors, this computer will receive larger segment. A parallel computation can be executed using OpenMP software with a compiler directive called "#pragma omp parallel", which divides the loop iterations. Segments with the amount that equals the amount of the processors can be computed simultaneously in one computer. The remaining segments allocated to the computer are started to be computed once one of the processors finishes computing of a segment. Once all these segments in these slave computers have finished computing, these feasible flight routes obtained have to be sent to the master computer to prepare for the computation of problem (4.3).

The amount of segments is proportional to the amount of processors, and more processors can enhance the efficiency of computation. Besides the high efficiency of this distributed computation, the feasible flight routes found by the division method in (4.3.2) are superior to those solved by CPLEX CP Optimizer when problem (4.3) is solved by these feasible flight routes. Comparison with CPLEX CP Optimizer is given in Section 4.4.

4.4 Computational experience

There are two flight schedules, 757 fleet and 737-100 fleet, appearing in [11], [10], and [26] and used in the evaluation; the details of these two schedules can be found in [10]. C++ is used to code the program, and

Distributed computation implementation
begin
MPI_Init(&argc,&argv);
MPI_Comm_rank(MPI_COMM,*computerid*)
GetSystemInfo(*SystemInfo*)
if(*computerid* = 0)
    begin
    receive *SystemInfo.dwNumberOfProcessors* from *computerid* ≠ 0
    divide the solution space based on the total number of processors and calculate
x^{e_i} and x^{s_i} defining each segment
    send x^{e_i} and x^{s_i} to each *computerid* ≠ 0
    #pragma omp parallel
        for(i=0;i< *SystemInfo.dwNumberOfProcessors*;i++)
            begin
            compute feasible flight routes of (4.1) in remaining segments by
CPLEX CP Optimizer or Dang's algorithm
            end
    receive feasible flight routes from each *computerid* ≠ 0
    end
else
    begin
    send *SystemInfo.dwNumberOfProcessors* to *computerid* = 0
    receive x^{e_i} and x^{s_i} from *computerid* = 0
    #pragma omp parallel
        for(i=0;i< *SystemInfo.dwNumberOfProcessors*;i++)
            begin
            compute feasible flight routes of (4.1) in received segments by CPLEX
CP Optimizer or Dang's algorithm
            end
    send feasible flight routes to *computerid* = 0
    end
end

Figure 4.6 Distributed computation pseudocode.

the linear programming in Dang's algorithm is solved by CPLEX Concert Technology. The version of the Cplex used is 12.6.1. First, problem (4.1) in Subsection 4.2.1 is solved for both two flight schedules by CPLEX CP Optimizer and Dang's algorithm, respectively. Then all these feasible flight routes are used to produce a solution of problem (4.3) in Subsection 4.2.2 considering the situation of grounding one aircraft from the beginning to the end of the schedule. Problem (4.3) is solved by CPLEX Optimizers's Concert Technology. Situations where each aircraft is grounded are all tested, and the comparison of the optimum values of problem (4.3) solved by CPLEX CP Optimizer and Dang's algorithm is presented.

In the implementation, a distributed computation network is developed by MPI. The feasible flight routes of problem (4.1) is solved by this network, where there are three different computers. One computer can build 16 threads at the same time to work, and the other two can build 2 threads. First, the results for 757 fleet are given. Two delay costs per minute are provided for the computation, and the comparison of the performances of three sets of feasible flight routes generated by CPLEX CP Optimizer with or without division and Dang's method with division to solve problem (4.3), respectively, is given. Then the results for 737-100 are provided. Two objectives are optimized: one aims at finding the minimum cost, and the other attempts to find the maximum profit. Comparison of these performances of two sets of partial feasible flight routes produced by CPLEX CP Optimizer with division and Dang's algorithm with division to solve problem (4.3), respectively, is presented.

4.4.1 757 fleet results

This fleet consists of 16 aircrafts that serve 42 flights between 13 cities a day and operates in a hub-and-spoke system. Thengvall [26] gives that the minimum turnaround time is 40 minutes, and the profit assigned to each flight route is $(1000 + 2 \times$ flight time $- 0.2 \times$ minutes delayed), so the delay cost is 0.2 per minute of delay. The solution of objective (4.3a) is the same as that of objective (4.3b) for this problem. So we only consider objective (4.3a). Argüello et al. [11] impose a delay cost of 20 per minute of delay on all delayed flights, and thus both these delay costs are used to solve problem (4.3).

Tables 4.2 and 4.3 are results of problem (4.3) under two different delay costs. Situations where aircrafts in the first columns of these two tables are grounded are all tested. These results are computed out by all feasible flight routes of problem (4.1). The result for delay cost of 0.2 in Table 4.2 contains combinations of feasible flight routes that delay much more time and have less cancellation than the result for delay cost of 20 in Table 4.3. It is obvious that less delay cost results in more delay time and less cancellation without considering other factors. For the grounding of aircraft 108 under delay cost 0.2, the resulting new schedule contains a flight leg that will delay 285 minutes. In practical solving an airline disruption problem, a flight leg delayed too much time causes great dissatisfaction of passengers and passengers lost in future. So usually the delay time is constrained within an upper bound, and all flight legs are canceled if their delay times exceed the upper

Table 4.2 Results of 757 fleet with delay cost of 0.2.

Aircraft	Delayed flights	Total delay minutes	Canceled flights	Swaps	Unaltered flight paths	Total cost
101	0	0	2 (170,203)	0	15	12 450
102	3 (1685,239,184)	396	0	5	10	79
103	3 (1685,239,184)	396	0	5	10	79
104	0	0	2 (711,712)	0	15	11 145
105	0	0	2 (176,197)	0	15	12 180
106	0	0	2 (192,189)	0	15	12 600
107	3 (1685,239,184)	396	0	5	10	79
108	1(184)	285	1(239)	2	13	6072
109	0	0	2 (240,195)	0	15	12 600
110	3 (1685,239,184)	396	0	5	10	79
111	1 (1685)	5	1 (196)	2	13	5746
112	3 (1685,239,184)	396	0	5	10	79
113	3 (1685,239,184)	396	0	5	10	79
114	0	0	2 (73,74)	0	15	12 555
115	0	0	2 (75,241)	0	15	12 555
116	0	0	2 (709,710)	0	15	11 145
Average	1.25	166.625	1.125	2.1875	12.875	6845.125

Table 4.3 Results of 757 fleet with delay cost of 20.

Aircraft	Delayed flights	Total delay minutes	Canceled flights	Swaps	Unaltered flight paths	Total cost
101	0	0	2 (170,203)	0	15	12 450
102	0	0	2 (1641,1640)	1	14	7110
103	0	0	2 (1641,1640)	1	14	7110
104	0	0	2 (711,712)	0	15	11 145
105	0	0	2 (176,197)	0	15	12 180
106	0	0	2 (192,189)	0	15	12 600
107	0	0	2 (1641,1640)	1	14	7110
108	1 (184)	285	1 (239)	1	13	11 715
109	0	0	2 (240,195)	0	15	12 600
110	0	0	2 (1641,1640)	1	14	7110
111	1 (1685)	5	1 (196)	2	13	5845
112	0	0	2 (1641,1640)	1	14	7110
113	0	0	2 (1641,1640)	1	14	7110
114	0	0	2 (73,74)	0	15	12 555
115	0	0	2 (75,241)	0	15	12 555
116	0	0	2 (709,710)	0	15	11 145
Average	0.125	18.125	1.875	0.5625	14.375	9840.625

bound. In this paper, we mainly discuss the universal situation without considering the impacts on passengers, and thus more feasible flight routes can be involved in computation of problem (4.3). More feasible flight routes mean more flexible in solving problem (4.3). If no solution can be found for problem (4.3) that is better than canceling the flight route original assigned to the grounding aircraft by more feasible flight routes, then a better solution for problem (4.3) cannot be found by less feasible flight routes. For situations of problem (4.3) that are hard to find a better solution, maybe we can find a better solution by providing much more feasible flight routes with unlimited delay time. The solution maybe would be useless in practice because it delays too much, but it is useful in the research of this airline disruption problem. Because for sets of feasible flight routes generated by different methods, comparison can only be made among those sets of feasible flight routes where at least one set of them can find a better solution of problem (4.3). If all these feasible flight route sets fail to solve a better solution of problem (4.3), then comparison is meaningless. Considering these reasons, the delay time is not limited to some discrete values in this chapter where as Thengvall [26] gives some constant times to delay. If needed, feasible flight routes with delay time in any specified value can be filtered out through a simple procedure.

For both two options of delay cost, each flight route containing only two flight legs and originating or/and terminating at a spoke station must be canceled if the original assigning aircraft is grounded from Tables 4.2 and 4.3. Because no feasible flight routes can be found for problem (4.3) except canceling the original flight route of the grounding aircraft. So these final solutions for the disruption problem are the same for grounding one of the aircrafts 101,104,105,106,109,114,115, and 116: just canceling the flight route that the grounding aircraft will fly, and the situation of grounding one of these aircrafts is not tested further. For each flight route that originates and terminates at the same hub station, the final results of reassigning aircrafts are the same when the original assigning aircraft is grounded. Thus these final solutions of grounding one of the aircrafts 102,103,107,110,112, and 113 are the same. Three sets of feasible flight routes of problem (4.1) are produced by CPLEX CP Optimizer with or without division and Dang's algorithm with division. Performances of these three sets of feasible flight routes in solving problem (4.3) that aircrafts 102, 108, and 111 are grounded are compared in the following.

Table 4.4 Performance of each thread using nearly average division.

	Thread	Start time (ms)	End time (ms)	Amount of solution	Duration per solution (ms)
Computer A	0	1000	4562	4	890.5
	1	1000	4843	5	768.6
Computer B	0	781	19 281	48	385.42
	1	781	1046	0	265
Computer C	0	982	3712	5	546
	1	982	44 616	110	396.67
	2	982	4758	7	539.43
	3	982	76 143	202	372.08
	4	982	237 947	849	279.11
	5	982	1294	0	312
	6	982	1950	1	968
	7	982	9079	17	476.29
	8	982	8970	17	469.88
	9	982	8517	16	470.94
	10	982	50 169	125	393.50
	11	982	18 642	39	452.82
	12	982	38 734	91	414.86
	13	982	126 375	399	314.27
	14	982	133 224	411	321.76
	15	982	503 350	1881	267.07
Total		781	503 350	4227	118.89

4.4.1.1 Performance of nearly average division

Since the 757 fleet schedule is relatively small, all feasible flight routes can be obtained within a reasonable time. Table 4.4 gives the computation performance of the distributed computation network when solving problem (4.1) of 757 fleet using Dang's algorithm combining nearly average division. The hardware of these computers is different, so their performances are not the same. Although the total solution space is divided into several segments containing more or less equal integer points, the amount of feasible flight routes differs greatly among these segments as in Table 4.4. No distribution of these feasible flight routes in the solution space can be obtained in advance, so nearly average division is not a best but universal method if all feasible flight routes are wanted to be obtained by this distributed computation method. More processors accelerate the computation of the problem.

Of course, CPLEX CP Optimizer also can solve problem (4.1) using nearly average division, it even costs much less time than Dang's algorithm. However, solutions obtained using Dang's algorithm are ordered in lexicographic order, whereas they are unordered using CPLEX CP Optimizer. There is no difference if all feasible flight routes are obtained and applied to compute problem (4.3), but the order affects the computation of the solution of problem (4.3) with partial feasible flight routes if all feasible flight routes cannot be obtained in a reasonable time. The importance of lexicographical order is explained in the following subsection.

4.4.1.2 Performance of initial seeds cluster division

Let initial seeds be original flight routes. Thus feasible flight routes in lexicographical order can be obtained sequentially using Dang's algorithm. Because the variable x_i in problem (4.1) marks the flight legs, they are rescheduled in increasing order of departure time. Therefore the earlier the solutions obtained from an original flight route, the less their deviation from the original flight route. Changes in these new obtained feasible flight routes are variations of flight legs in the back of the original flight route at the beginning of the computation using Dang's algorithm. These variations of flight legs spread from the back of the original flight route to the front of it along with the computation process, and thus these later obtained feasible flight routes are more deviated from the original one. One objective of solving this airline disruption problem is finding an aircraft reassignment that is as less deviated as possible from the original assignment. The conclusion that most optimum solutions of problem (4.3) are computed out with those feasible flight routes that are less deviated from original ones can be drawn from the following comparison with CPLEX CP Optimizer, and only with a small percentage of total feasible flight routes produced by Dang's algorithm, the optimum solution of problem (4.3) can be found.

Table 4.5 shows the performance of Dang's algorithm using initial seeds cluster division to solve problem (4.1). There are 16 aircrafts in 757 fleet, so the amount of original flight routes is 16, so that 16 clusters are defined based on these center seeds. The parameter cluster size controls the amount of partial feasible flight routes produced in each cluster. However, the amount of total feasible flight routes is less than 16 × cluster size, because the amount of partial feasible flight routes in some clusters is less than the cluster size. The transformation procedure in Fig. 4.2 eliminates some feasible flight routes that exceed the departure curfew time of stations. Tables 4.6 and 4.7 show the performances of CPLEX CP Optimizer with

Table 4.5 Performance of Dang's algorithm using initial seeds cluster division.

Cluster size	Amount of solutions	Remaining solutions after transformation	Duration (ms)	Duration per solution (ms)
10	159	140	3806	23.94
20	312	272	7355	23.57
30	455	402	10 951	24.07
40	589	519	14 805	25.14
50	714	631	18 520	25.94
60	836	737	22 105	26.44
70	951	837	25 818	27.15
80	1062	938	29 110	27.41
90	1155	1018	32 292	27.96
100	1237	1091	35 162	28.43
150	1615	1423	47 096	29.16
200	1930	1717	60 934	31.57

Table 4.6 Performance of CPLEX CP Optimizer using initial seeds cluster division.

Cluster size	Amount of solutions	Remaining solutions after transformation	Duration (ms)	Duration per solution (ms)
10	159	156	210	1.32
20	312	296	230	0.74
30	455	417	250	0.55
40	589	540	250	0.42
50	714	658	270	0.38
60	836	771	290	0.35
70	951	874	300	0.32
80	1062	969	330	0.31
90	1155	1052	360	0.31
100	1237	1114	360	0.29
150	1615	1446	410	0.25
200	1930	1728	450	0.23

and without initial seeds cluster division. No cluster is defined for CPLEX CP Optimizer without division, so there is no column for cluster size. The amounts of partial feasible flight routes in each cluster in Tables 4.5 and 4.6 are adopted in CPLEX CP Optimizer without division to produce partial feasible flight routes in the same amount. The numbers of partial feasible flight routes produced by Dang's algorithm, CPLEX CP Optimizer with division, and CPLEX CP Optimizer without division after transformation

Table 4.7 Performance of CPLEX CP Optimizer without division.

Amount of solutions	Remaining solutions after transformation	Duration (ms)	Duration per solution (ms)
159	149	20	0.126
312	269	30	0.096
455	366	40	0.088
589	494	50	0.085
714	593	50	0.07
836	705	60	0.072
951	786	70	0.074
1062	871	80	0.075
1155	958	80	0.069
1237	1034	90	0.073
1615	1338	130	0.08
1930	1593	170	0.088

in each cluster size are different due to differences among these three sets of partial feasible flight routes generated by the three approaches. These little differences in amounts of partial feasible flight routes that participate in computation problem (4.3) do not affect the comparison results. There are two lexicographical constraints to limit the feasible flight routes generated within the bounds that define each segment and a quadratic constraint to preclude the seed. So it costs more time for CPLEX CP Optimizer using initial seeds cluster division than for CPLEX CP Optimizer without division. Both CPLEX CP Optimizer using division or not cost much less time than Dang's algorithm, but performances of these partial feasible flight routes produced using CPLEX CP Optimizer do worse than those generated by Dang's algorithm as illustrated in the following figures.

Figs. 4.7–4.9 reveal the performances of partial feasible flight routes of problem (4.1) generated by CPLEX CP Optimizer and Dang's algorithm to solve problem (4.3). Figs. 4.7(a) and 4.7(b) are comparisons of the two methods when grounding aircraft 102, and the situation is the same for grounding one of the aircrafts 103, 107, 110, 112, and 113 from Tables 4.2 and 4.3. Figs. 4.8 and 4.9 are comparisons of the three approaches when grounding aircraft 108 and 111, respectively. The results for grounding one of aircrafts 101,104,105,106,109,114,115, and 116 are all the same and cancel the flight route original assigned to the grounding aircraft. So no better solution can be found for problem (4.3) by all of these approaches when grounding one of these aircrafts.

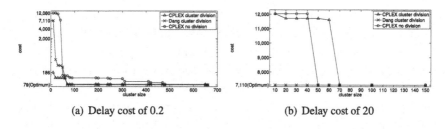

Figure 4.7 Comparison of CPLEX and Dang grounding aircraft 102.

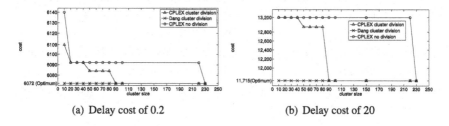

Figure 4.8 Comparison of CPLEX and Dang grounding aircraft 108.

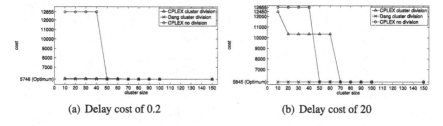

Figure 4.9 Comparison of CPLEX and Dang grounding aircraft 111.

The maximal minimum cluster size needed for partial feasible flight routes produced by Dang's algorithm with initial seed division to find the optimum of problem (4.3) is 70, whereas the value is 660 for partial feasible flight routes produced by CPLEX CP Optimizer with division and 480 for partial feasible flight routes produced by CPLEX CP Optimizer without division in situations of grounding one of aircrafts 102, 103, 107, 110, 112, and 113 under delay cost of 0.2 in Fig. 4.7(a). For the minimal minimum cluster size needed for partial feasible flight routes produced by Dang's algorithm with division is only 10, but both partial feasible flight routes produced by CPLEX CP Optimizer with or without division need cluster size more than 50 in Figs. 4.7(b), 4.8, and 4.9.

From this comparison it follows that CPLEX CP Optimizer without division needs huge amount of partial feasible flight routes to find the optimum for problem (4.3). CPLEX CP Optimizer using initial seeds cluster division performs better than no division for most of these situations but costs more time. Dang's algorithm needs the fewest partial feasible flight routes to hit the optimum at the cost of consuming more time than CPLEX CP Optimizer. So determination of which method is better cannot be made to this 757 fleet problem. Besides, this 757 fleet is a small problem because all feasible flight routes can be obtained in less than one second using CPLEX CP Optimizer, and the optimum for problem (4.3) can be found using all feasible flight routes. So CPLEX CP Optimizer seems outperform Dang's algorithm in this small 757 fleet problem. However, from the comparison of the following 737-100 fleet, CPLEX CP Optimizer fails to find a solution even when hundreds of thousand partial feasible flight routes are provided.

4.4.2 737-100 fleet results

This fleet consists of 27 aircrafts in the same fleet that serve 162 flights between 30 cities a day and also operates in a hub-and spoke system. Thengvall [26] gives that the minimum turnaround time is 25 minutes, the bonus per flight leg covered is 300, and the delay cost per minute is 5. In Section 4.2.2, we mentioned that the bonus per flight leg covered encourages the route to be less deviated from the original flight route. This bonus per flight leg covered is involved in calculating the profit p_j of each feasible flight route. It does not participate in calculating cost of each feasible flight route, because there is no cost per flight leg not covered that charges flight legs that are not original belonging to the flight route. So results for objectives (4.3a) and (4.3b) are different, and both these objectives are tested.

This fleet is a medium-size fleet and costs 1.27×10^{45} s, which is about 4×10^{37} years, in the worst case to find all feasible flight routes out by CPLEX CP Optimizer mentioned in Section 4.3. It takes about half a month to generate three hundred million feasible flight routes, but it still goes on generating, so the generation is canceled for it costs too much time. Since it is impossible to produce all feasible flight routes in a short time, problem (4.3) is solved just by the partial feasible flight routes. So the solution space of this problem is segmented by initial seeds cluster division rather than by nearly average division, and CPLEX CP Optimizer

Table 4.8 Results of 737-100 fleet for objective (4.3a) in cluster size 5000.

Aircraft	Delayed flights	Total delay minutes	Can-celed flights	Swaps	Unaltered flight paths	Total cost
201	6	1012	2	8	20	9395
202	8	130	4	9	19	11 705
203	3	494	4	3	24	13 225
204	0	0	5	2	24	12 090
207	0	0	5	1	24	10 854
208	0	0	6	0	26	13 605
209	7	885	3	11	19	11 475
211	1	30	6	3	23	12 900
212	4	278	5	10	19	13 675
213	9	150	3	10	19	10 215
214	6	117	4	8	21	10 500
215	5	116	4	4	23	11 320
231	9	150	3	11	18	8160
232	3	450	4	9	21	11 220
233	6	117	3	6	21	8205
234	2	109	4	8	21	9245
235	3	494	4	5	23	12 910
239	9	366	3	13	18	9915
241	10	533	2	13	17	7120
242	8	130	3	12	19	10 025
243	5	617	3	10	18	9205
245	7	137	2	10	20	5695
246	7	91	4	8	21	8660
248	0	0	4	1	25	8160
251	4	108	3	7	21	9030
252	6	117	3	7	20	8385
255	0	0	6	0	26	13 080
Average	4.74	245.593	3.778	7	21.111	10 369.07

and Dang's algorithm are used to produce partial feasible flight routes of problem (4.1).

Tables 4.8 and 4.9 are the results for objectives (4.3a) and (4.3b) using CPLEX Optimizers's Concert Technology with partial feasible flight routes produced by Dang's algorithm with division cluster size 5000. Situations where each aircraft is grounded are all considered. The delay time is also not limited to some discrete values.

Table 4.9 Results of 737-100 fleet for objective (4.3b) in cluster size 5000.

Aircraft	De-layed flights	Total delay minutes	Can-celed flights	Swaps	Unaltered flight paths	Total cost
201	5	1115	2	4	23	423 687
202	3	511	4	2	24	420 943
203	3	686	4	2	24	420 347
204	0	0	7	0	26	421 905
207	0	0	5	1	24	424 116
208	0	0	6	0	26	421 980
209	3	206	5	2	24	421 387
211	0	0	7	0	26	419 865
212	2	412	5	1	25	420 480
213	3	492	3	2	24	422 894
214	3	381	4	2	24	422 847
215	2	392	4	2	24	422 477
231	5	474	3	3	23	424 019
232	2	602	4	2	24	423 065
233	3	354	3	3	23	424 751
234	2	145	4	1	25	425 622
235	3	516	4	3	24	420 310
239	2	779	3	3	23	422 905
241	0	0	5	1	24	424 506
242	3	553	3	2	24	422 719
243	3	732	3	2	24	424 359
245	3	407	2	3	23	427 396
246	3	335	4	2	24	424 790
248	0	0	4	1	25	427 101
251	3	242	3	2	24	425 371
252	3	347	3	3	23	424 606
255	0	0	6	0	26	422 505
Average	2.185	358.556	4.074	1.815	24.185	423 220.48

4.4.2.1 Performance of initial seeds cluster division

In this subsection, the partial feasible flight routes with their amount controlled by cluster size are produced by initial seeds cluster division. CPLEX CP Optimizer without division is no longer applied here for its poor performance. Both CPLEX CP Optimizer and Dang's algorithm combining initial seeds cluster division are used to generate two sets of partial feasible

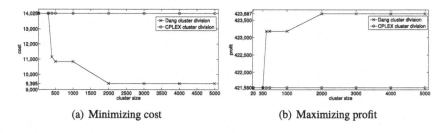

(a) Minimizing cost   (b) Maximizing profit

Figure 4.10 Comparison of CPLEX and Dang grounding aircraft 201.

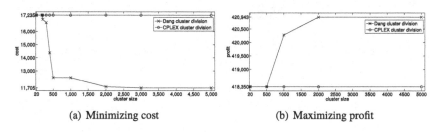

(a) Minimizing cost   (b) Maximizing profit

Figure 4.11 Comparison of CPLEX and Dang grounding aircraft 202.

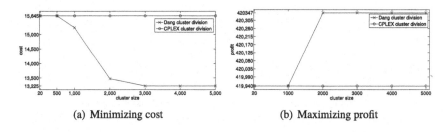

(a) Minimizing cost   (b) Maximizing profit

Figure 4.12 Comparison of CPLEX and Dang grounding aircraft 203.

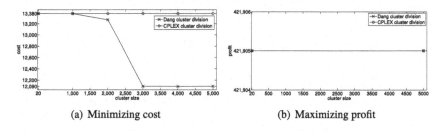

(a) Minimizing cost   (b) Maximizing profit

Figure 4.13 Comparison of CPLEX and Dang grounding aircraft 204.

flight routes. The results of problem (4.3) computed by these two sets of partial feasible flight routes are compared in Figs. 4.10–4.36.

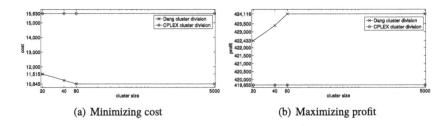

(a) Minimizing cost   (b) Maximizing profit

Figure 4.14 Comparison of CPLEX and Dang grounding aircraft 207.

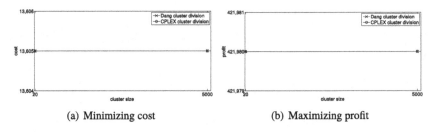

(a) Minimizing cost   (b) Maximizing profit

Figure 4.15 Comparison of CPLEX and Dang grounding aircraft 208.

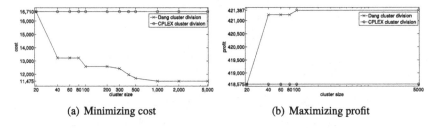

(a) Minimizing cost   (b) Maximizing profit

Figure 4.16 Comparison of CPLEX and Dang grounding aircraft 209.

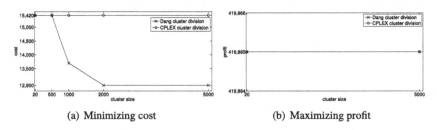

(a) Minimizing cost   (b) Maximizing profit

Figure 4.17 Comparison of CPLEX and Dang grounding aircraft 211.

15 out of total 27 cases using partial feasible flight routes produced by Dang's algorithm combining initial seeds cluster division only need cluster with a size less than or equal to 100 to find a solution for objective (4.3a),

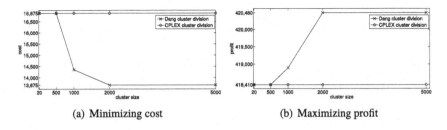

(a) Minimizing cost   (b) Maximizing profit

Figure 4.18 Comparison of CPLEX and Dang grounding aircraft 212.

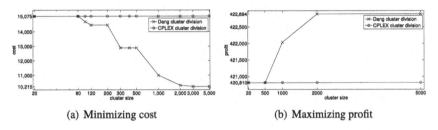

(a) Minimizing cost   (b) Maximizing profit

Figure 4.19 Comparison of CPLEX and Dang grounding aircraft 213.

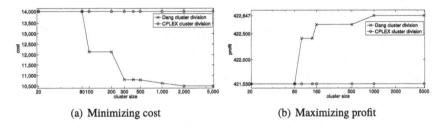

(a) Minimizing cost   (b) Maximizing profit

Figure 4.20 Comparison of CPLEX and Dang grounding aircraft 214.

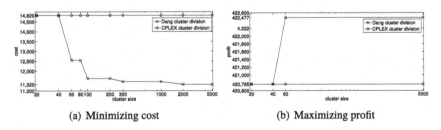

(a) Minimizing cost   (b) Maximizing profit

Figure 4.21 Comparison of CPLEX and Dang grounding aircraft 215.

which is better than canceling flight route original assigned to the grounding aircraft. A larger cluster size is required for these remaining cases to
find a better solution by Dang's algorithm. For objective (4.3b), 14 out of

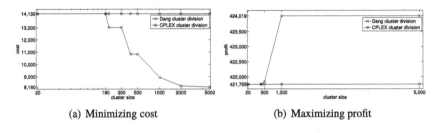

(a) Minimizing cost   (b) Maximizing profit

Figure 4.22 Comparison of CPLEX and Dang grounding aircraft 231.

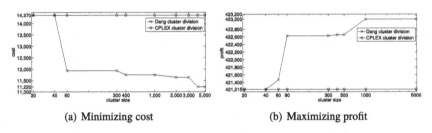

(a) Minimizing cost   (b) Maximizing profit

Figure 4.23 Comparison of CPLEX and Dang grounding aircraft 232.

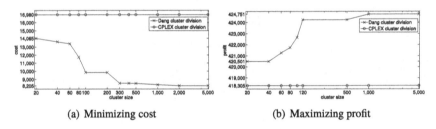

(a) Minimizing cost   (b) Maximizing profit

Figure 4.24 Comparison of CPLEX and Dang grounding aircraft 233.

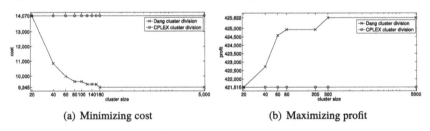

(a) Minimizing cost   (b) Maximizing profit

Figure 4.25 Comparison of CPLEX and Dang grounding aircraft 234.

total 27 cases only need cluster of size less than or equal to 100 to find a better solution. These remaining cases also need a larger cluster size to find a better solution.

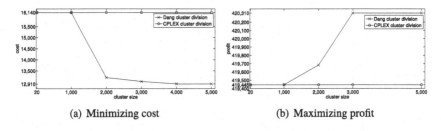

(a) Minimizing cost

(b) Maximizing profit

Figure 4.26 Comparison of CPLEX and Dang grounding aircraft 235.

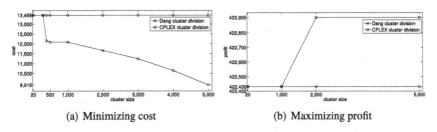

(a) Minimizing cost

(b) Maximizing profit

Figure 4.27 Comparison of CPLEX and Dang grounding aircraft 239.

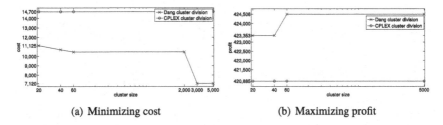

(a) Minimizing cost

(b) Maximizing profit

Figure 4.28 Comparison of CPLEX and Dang grounding aircraft 241.

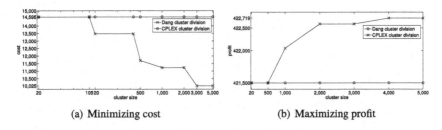

(a) Minimizing cost

(b) Maximizing profit

Figure 4.29 Comparison of CPLEX and Dang grounding aircraft 242.

All cases just need a cluster of size no greater than 2000 to find a better solution by Dang's algorithm except situations in Figs. 4.13(b), 4.15, 4.17(b), and 4.36 where both Dang's algorithm and CPLEX CP Optimizer

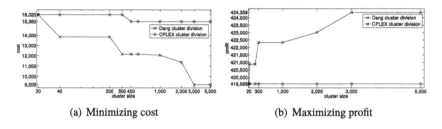

(a) Minimizing cost   (b) Maximizing profit

Figure 4.30 Comparison of CPLEX and Dang grounding aircraft 243.

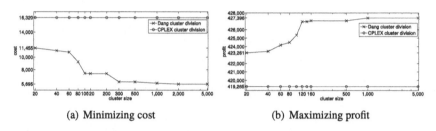

(a) Minimizing cost   (b) Maximizing profit

Figure 4.31 Comparison of CPLEX and Dang grounding aircraft 245.

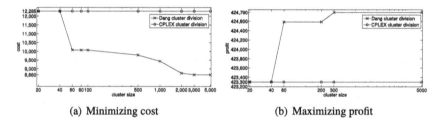

(a) Minimizing cost   (b) Maximizing profit

Figure 4.32 Comparison of CPLEX and Dang grounding aircraft 246.

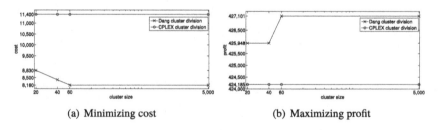

(a) Minimizing cost   (b) Maximizing profit

Figure 4.33 Comparison of CPLEX and Dang grounding aircraft 248.

fail to find a better solution within cluster size 5000. Maybe there is no better solution than canceling all flights original assigned to the grounding aircrafts for situations in Figs. 4.13(b), 4.15, 4.17(b), and 4.36. For all

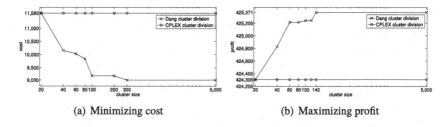

(a) Minimizing cost   (b) Maximizing profit

Figure 4.34 Comparison of CPLEX and Dang grounding aircraft 251.

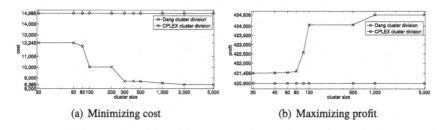

(a) Minimizing cost   (b) Maximizing profit

Figure 4.35 Comparison of CPLEX and Dang grounding aircraft 252.

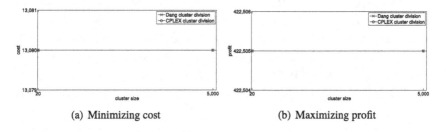

(a) Minimizing cost   (b) Maximizing profit

Figure 4.36 Comparison of CPLEX and Dang grounding aircraft 255.

results computed using partial feasible flight routes within the cluster size 5000 generated by CPLEX CP Optimizer, combining initial seed division fails to obtain a better solution except for the situation of grounding aircraft 243 and minimizing the cost in Fig. 4.30(a).

The cluster size is increased up to 20000 for CPLEX CP Optimizer to find a better solution of problem (4.3), and these solutions of problem (4.3) computed by these partial feasible flights routes are compared to these solutions generated by Dang's algorithm of a relatively small size of cluster in Tables 4.10 and 4.11. The situations where each aircraft is grounded are all considered, but situations where CPLEX CP Optimizer combining initial seeds cluster division of cluster size 20000 still cannot find a solution better than canceling the flight route original assigned to the grounding aircraft are omitted here.

Table 4.10 is the comparison result of objective (4.3a) by partial feasible flight routes generated by the two methods. Only in situations of grounding aircrafts in Table 4.10, CPLEX CP Optimizer can find a solution better than canceling flight route original assigned to the grounding aircraft. That is to say, for the situations of grounding aircrafts not appearing in the table, more feasible flight routes are required to find a better solution using CPLEX CP Optimizer, whereas much less partial feasible flight routes are required by Dang's algorithm. Situations where grounding one of aircrafts 207, 209, 233, 243, and 252 need much less partial feasible flight routes produced by Dang's algorithm and less total duration of computation to find a solution of problem (4.3) compared to partial feasible flight routes produced by CPLEX CP Optimizer. The final solution values obtained by partial feasible flight routes produced by Dang's algorithm are also better than those obtained by partial feasible flight routes produced by CPLEX CP Optimizer in these situations. The total duration of using Dang's algorithm is a little larger than that of using CPLEX CP Optimizer in situations of grounding one of aircrafts 245 and 246. However, the need for the amount of partial feasible flight routes produced using Dang's algorithm is much smaller than the need for the amount of partial feasible flight routes produced using CPLEX CP Optimizer. For situations of grounding one of aircrafts 202, 214, 231, and 242, more total duration of computation is needed by CPLEX CP Optimizer than by Dang's algorithm.

Table 4.11 is the comparison result of maximizing profit by partial feasible flight routes generated through the two methods. Only in situation of grounding aircraft 245 can CPLEX CP Optimizer find a solution better than canceling the flight route original assigned to aircraft 245. CPLEX CP Optimizer seems to lack success to produce feasible flight routes that could constitute a new scheduler for the remaining aircrafts with a total profit larger than the original assignment without the flight route of the grounding aircraft in solving objective (4.3b) that aims to maximize profit.

The generation of partial feasible flight routes for problem (4.1) using Dang's algorithm is not as efficient as CPLEX CP Optimizer, but only a few of partial feasible flight routes produced by Dang's algorithm are required to solve a better solution of problem (4.3), whereas a huge amount of partial feasible flight routes produced by CPLEX CP Optimizer is required to find a better solution.

Table 4.10 Comparison on minimizing cost of CPLEX and Dang.

Aircraft	Method	Cluster size	Solution size of (4.1)	Solution size of (4.1) after transformation	Cost	Duration (ms)		Total duration (ms)
						Problem (4.1)	Problem (4.3)	
202	Dang	400	10800	2806	14375	6060734	3168	6063902
	CPLEX	20000	540000	86886	14545	756851	102896	859747
207	Dang	20	540	271	11515	286194	411	286605
	CPLEX	20000	540000	86886	14480	756851	102523	859374
209	Dang	40	1080	442	13235	577985	592	578577
	CPLEX	20000	540000	86886	15685	756851	106127	862978
214	Dang	100	2700	923	12135	1484797	1045	1485842
	CPLEX	20000	540000	86886	13955	756851	103459	860310
231	Dang	200	5400	1536	13020	2907844	1721	2909565
	CPLEX	20000	540000	86886	13040	756851	103661	860512
233	Dang	20	540	271	14065	286194	406	286600
	CPLEX	20000	540000	86886	15920	756851	103587	860438
242	Dang	120	3240	1030	13475	1766328	1240	1767568
	CPLEX	20000	540000	86886	13825	756851	103740	860591
243	Dang	40	1080	442	13840	577985	577	578562
	CPLEX	20000	540000	86886	15150	756851	103632	860483
245	Dang	60	1620	617	10785	866719	781	867500
	CPLEX	20000	540000	86886	10700	756851	105622	862473
246	Dang	60	1620	617	10080	866719	780	867499
	CPLEX	20000	540000	86886	11755	756851	101401	858252
252	Dang	20	540	271	12245	286194	406	286600
	CPLEX	20000	540000	86886	13565	756851	110604	867455

Table **4.11** Comparison on maximizing profit of CPLEX and Dang.

Aircraft	Method	Cluster size	Solution size of (4.1)	Solution size of (4.1) after transformation	Profit	Duration/ms Problem (4.1)	Duration/ms Problem (4.3)	Total duration (ms)
245	Dang	20	540	271	423261	286194	421	286615
	CPLEX	20000	540000	86886	422615	756851	105020	861871

For the time-consuming of Dang's algorithm, many improvements can be made. For this 737-100 fleet, there exist 27 original flight routes, so the amount of segments is 27 based on these original flight routes. However, there are only 20 processors in our distributed implementation, less than the amount of segments. Only 20 segments can be computed simultaneously, and the remaining 7 segments have to wait till one of the previous 20 segments finishes computing. So the efficiency can be improved by providing more computation processors. The amount of computation processors is recommended to be multiple of the amount of original flight routes because each segment can be divided again by nearly average division in Subsection 4.3.1.

Original schedules are usually determined a couple of days before the execution of schedules [2], so partial feasible flight routes of problem (4.1) can be started to generate once the original schedules are ready. There is no need to wait till a disruption happened that feasible flight routes are started to produce. Once a disruption happened, all these partial feasible flight routes get ready to find the solution of problem (4.3). So the computation duration of problem (4.1) can be ignored for the timing of the total duration of solving this airline disruption problem. Only the computation duration of problem (4.3) is counted for the timing of the total duration. Based on this point of view, Dang's algorithm outperforms CPLEX CP Optimizer for much less partial feasible flight routes produced by Dang's algorithm are required than those produced by CPLEX CP Optimizer to solve a better solution of problem (4.3). With less partial feasible flight routes, the solution of problem (4.3) can be computed out quickly, whereas much longer duration is needed with more partial feasible flight routes from Tables 4.10 and 4.11.

If the amount of partial feasible flight routes is not enough to obtain a better solution of problem (4.3), then the generation of feasible flight routes could be continued from the last obtained feasible flight route in each segment by Dang's algorithm.

4.5 Conclusion

In this chapter, we presented a procedure to compute the airline disruption problem [27]. The procedure includes two subproblems, which are solving feasible flight routes generation in Subsection 4.2.1 and aircrafts reassignment in Subsection 4.2.2. Here we mainly discussed the subproblem of feasible flight routes generation. For a long-haul airline disruption problem, it is impossible to produce all feasible flight routes in a short time, so the solution space for this subproblem is divided into several segments using two division methods. The first division method is to divide the whole solution space into more or less equal segments, and each segment can be computed simultaneously in an individual processor. So providing sufficient processors, the solution space can be divided into segments that contain a small amount of feasible flight routes. For the situation of lack of processors, a second division method is proposed, which divides the solution space into clusters. Each cluster is contained in a segment defined by two bound points that lie in the middle of two continuous original flight routes in lexicographical order. Considering the hardness to obtain all feasible flight routes in each segment, only partial feasible flight routes are given to compute problem (4.3), and the amount of these partial feasible flight routes is controlled by the parameter cluster size. A distributed computation based on Dang's algorithm and these two division methods is applied to produce partial feasible flight routes. With the computation experience of two flight schedules, much less partial feasible flight routes produced by Dang's algorithm are required to obtain a better solution for problem (4.3) than those produced by CPLEX CP Optimizer. For the long-haul airline schedule of the second flight schedule, only one case of total 54 could obtain a better solution for problem (4.3) with the help of partial feasible flight routes produced by CPLEX CP Optimizer within the cluster size 5000 contrasting to 48 cases of total 54 by partial feasible flight routes produced by Dang's algorithm within the cluster size 2000. When the size of cluster is increased up to 20000 for CPLEX CP Optimizer to produce partial feasible flight routes, 12 cases of total 54 find a better solution of problem (4.3), which is still much less than 48 cases of total 54 by Dang's algorithm. The solution of problem (4.3) is more and more close to the optimum of problem (4.3) as the cluster size is increasing, but these flight routes of the solution are more deviated from original flight routes. For long-haul airline problems that are impossible to obtain all feasible solutions in a short time, a properly chosen cluster size can obtain a solution of problem (4.3) that is at least very close to the optimum and less deviated from original flight routes by Dang's algorithm combining initial seeds cluster division. Although only results for situations of grounding one aircraft are given in this chapter, these results for situations of grounding more than one aircraft are more or less the same. So these results are omitted here due to the limitation of the length of this chapter. For the time-consuming of Dang's algorithm to generate feasible flight routes, improvement can be made by providing sufficient computation processors. In a practical application, the procedure of generation of feasible flight routes can be conducted as soon as the original schedule is decided. These feasible flight routes can be immediately used to solve problem (4.3) when a disruption happens, and the duration of generating these feasible flight routes can be ignored for timing the total duration of computing this airline disruption problem. Only the computation duration of problem (4.3) is counted in the total duration of solving this airline disruption problem, and Dang's algorithm outperforms CPLEX CP Optimizer for it can provide much less partial feasible flight routes, which costs much less time in computation than CPLEX CP Optimizer to obtain a better solution of problem (4.3).

References

[1] Chuangyin Dang, Yinyu Ye, A fixed point iterative approach to integer programming and its distributed computation, Fixed Point Theory and Applications 2015 (1) (2015) 1–15.
[2] Jens Clausen, Allan Larsen, Jesper Larsen, Natalia J. Rezanova, Disruption management in the airline industry concepts, models and methods, Computers and Operations Research 37 (May 2010) 809–821.
[3] Chi-Ruey Jeng, Multi-fleet Airline Schedule Disruption Management – Using an Inequality-based Multiobjective Genetic Algorithm, Ph.d. Dissertation, Department of Transportation and Communication Management Science, National Cheng-Kung University, Taiwan, 2009.
[4] Association of European Airline. European airline punctuality report, 2008.
[5] Dušan Teodorović, Slobodan Guberinić, Optimal dispatching strategy on an airline network after a schedule perturbation, European Journal of Operational Research 15 (2) (1984) 178–182.
[6] Dušan Teodorović, Goran Stojković, Model for operational daily airline scheduling, Transportation Planning and Technology 14 (4) (1990) 273–285.
[7] Dušan Teodorović, Goran Stojković, Model to reduce airline schedule disturbances, Journal of Transportation Engineering 121 (4) (1995) 324–331.
[8] Ahmad I.Z. Jarrah, Gang Yu, Nirup Krishnamurthy, Ananda Rakshit, A decision support framework for airline flight cancellations and delays, Transportation Science 27 (3) (1993) 266–280.
[9] Yu. Gang, An optimization model for airlines' irregular operations control, in: Proceedings of the International Symposium on Operations Research with Applications in Engineering, Technology, and Management, Beijing, China, 1995.
[10] Michael F. Argüello, Framework for exact solutions and heuristics for approximate solutions to airlines' irregular operations control aircraft routing problem, Ph.D. dissertation, Department of Mechanical Engineering, University of Texas, Austin, 1997.
[11] Michael F. Argüello, Jonathan F. Bard, Gang Yu, A grasp for aircraft routing in response to groundings and delays, Journal of Combinatorial Optimization 1 (1997) 211–228.
[12] Michael F. Argüello, Jonathan F. Bard, Gang Yu, Models and methods for managing airline irregular operations, in: Operations Research in the Airline Industry, in: International Series in Operations Research & Management Science, vol. 9, Springer US, 1998, pp. 1–45.

[13] Ladislav Lettovský, Airline operations recovery: an optimization approach, Ph.D. dissertation, Department of Industrial and Systems Engineering, Georgia Institute of Technology, Atlanta, GA, 1997.

[14] Benjamin G. Thengvall, Jonathan F. Bard, Gang Yu, Balancing user preferences for aircraft schedule recovery during irregular operations, IIE Transactions 32 (2000) 181–193.

[15] Jonathan Bard, Gang Yu, Michael Argüello, Optimizing aircraft routings in response to groundings and delays, IIE Transactions 33 (2001) 931–947.

[16] Michael Løve, Kim Riis Sørensen, Jesper Larsen, Jens Clausen, Disruption management for an airline – rescheduling of aircraft, in: Stefano Cagnoni, Jens Gottlieb, Emma Hart, Martin Middendorf, Günther Raidl (Eds.), Applications of Evolutionary Computing, in: Lecture Notes in Computer Science, vol. 2279, Springer, Berlin/Heidelberg, 2002, pp. 289–300.

[17] Jay M. Rosenberger, Ellis L. Johnson, George L. Nemhauser, Rerouting aircraft for airline recovery, Transportation Science 37 (2003) 408–421.

[18] Tobias Andersson, Peter Värbrand, The flight perturbation problem, Transportation Planning and Technology 27 (2) (2004) 91–117.

[19] Shangyao Yan, Chin-Hui Tang, Chong-Lan Shieh, A simulation framework for evaluating airline temporary schedule adjustments following incidents, Transportation Planning and Technology 28 (3) (2005) 189–211.

[20] Niklas Kohl, Allan Larsen, Jesper Larsen, Alex Ross, Sergey Tiourine, Airline disruption management – perspectives, experiences and outlook, Journal of Air Transport Management 13 (3) (2007) 149–162.

[21] Cheng-Lung Wu, Airline Operations and Delay Management – Insignts from Airline Economics, Networks and Strategic Schedule Planning, Ashgate, 2010.

[22] Massoud Bazargan, Airline Operations and Scheduling, Ashgate, 2004.

[23] Dušan Teodorović, Airline Operations Research (Transportation Studies), Routledge, 1988.

[24] Ahmed Abdelghany, Khaled Abdelghany, Modeling Applications in the Airline Industry, Ashgate, 2010.

[25] Gang Yu (ed.), Operations Research in the Airline Industry (International Series in Operations Research & Management Science), Springer, 1997.

[26] Benjamin G. Thengvall, Models and Solution Techniques for the Aircraft Schedule Recovery Problem, Ph.D. dissertation, Graduate Program in Operations Research and Industrial Engineering, University of Texas, Austin, 1999.

[27] Zhengtian Wu, Benchi Li, Chuangyin Dang, Fuyuan Hu, Qixin Zhu, Baochuan Fu, Solving long haul airline disruption problem caused by groundings using a distributed fixed-point computational approach to integer programming, Neurocomputing 269 (2017) 232–255.

CHAPTER 5

Solving multiple fleet airline disruption problems using a distributed-computation approach

Generally, traveling by aircrafts is the most convenient and time–saving way for long-distance transportation. Therefore the punctuality of airlines is significant to the airline carriers because unpunctuality always causes great inconvenience to the passengers and brings losses to airlines. To pursue a good punctuality, the operation of an airline requires an elaborate construction of an airline schedule planning, which includes flight scheduling, fleet assignment, aircraft routing, and crew scheduling. The ideal execution of the airline operation is identical to the flight schedules [1], but it is seldom operated as original plan due to unexpected disruptions such as aircraft breakdowns, crew absences, severe weather conditions, and air traffic and airport restrictions. Thus aircrafts, crews, and passengers are affected by disruptions. A disruptions can cause misconnect, rest, and duty problems to the schedules of the crews [2]. The passengers may miss their connecting flights or other transportation, which leads to economy losses. For the airlines, the ground time they spend at airports between two consecutive flights is minimized to maximize the utilization of each aircraft. Therefore any disruption can cause downstream impact on the subsequent flights of the disrupted aircrafts since aircrafts and crews may not be available at the scheduled time of the downstream flights [2,3]. The consequential disruption of the downstream flights is called delay propagation, and it affects the airline operations much more than the initial delay. The disruptions not only cause great inconvenience to the passengers but also bring losses to airlines. For a typical airline, losses caused by the disruptions approximately take up 10% of its scheduled revenue according to [4].

In mainland China the punctuality performance of all the airlines in 2010 is reported in [5]. There are 1888 thousands scheduled flights in the major airline companies. 1431 thousands flights perform as scheduled, but 457 thousands flights or 24.2% of the total flights perform irregularly. For

Integer Optimization and Its Computation in Emergency Management
https://doi.org/10.1016/B978-0-32-395203-3.00010-1
85

the middle and small airlines, of the total 260 thousand scheduled flights, there are 81 thousand flights operated irregularly. The percentage is up to 31.2%. The total losses caused by the irregularities reaches CNY2.1 billion in 2002, and it is estimated to increase to CNY7.6 billion in 2020 [6]. The general manager from China Eastern Air Holding Co. points out that the direct operating cost of China Eastern Airlines is CNY1,000 for each minute delay in 2011, and the cost does not cover the compensation and service to the passengers [7].

In the Europe, it is revealed by European Organization for the Safety of Air Navigation (EUROCONTROL) [8] and *Central Office for Delay Analysis (CODA)*[1] [9] that 18% of the flights are delayed on arrival by more than 15 minutes in 2011. EUROCONTROL [8] shows the estimated cost of ATFM delays decreases from C2.2 billion in 2010 to C1.45 billion in 2011, but it is still higher than that in 2009 by 21%. In the United States, the average percentage of the punctual flights from September 1987 to December 2003 is 78.9% [3]. The lowest point is only 72.6%, which happened in December 2000 [3]. The U.S. reports that the direct aircraft operating cost per minute is $65.19 in 2010, and it increases by 6% versus 2009 [10]. The total delay costs come up to $6,475 millions in 2010.

Since the disruptions can result in huge revenue losses to airlines inevitably, the punctuality performance affects the profit of the airlines. However, the research in [11] shows that most airlines with high punctuality rates appear to be more profitable than those with low punctuality rates. When the airline disruptions happen, producing a recovery plan can reduce the losses and increase the punctuality.

It is a complex task to produce recovery plans since many resources such as crews, aircrafts, and passengers have to be reassigned, and this problem has attracted researchers' interests since the 1980s. A branch-and-bound method is used to produce the recovery plan to minimize total passenger delays [12], and the work is further extended in [13,14]. All of them model the airline disruption problem as a connection network, which is also adopted in [15] and [16]. In [17,18] the time-line network is used to formulate the problem. Other papers [19,20] propose a time-band network to model the airline disruption problem. A framework for the integrated recovery that considers aircraft, crew, and passenger together is proposed in [21]. In terms of solution methods, the research in [18,19] applies heuristics

---

[1] CODA is a service of EUROCONTROL.

to solve the airline disruption problems based on the models they formulated, whereas the majority of the other researchers mentioned before use integer programming. An introduction to airline disruption management is given in [22]. An inequality-based multiobjective genetic algorithm (MMGA) is capable of solving multiple objective airline disruption problems is developed in [23]. For more theoretical descriptions and comparisons regarding the airline disruption management, we refer to the review in [1]. Some new methods and theory [24–27] can also be used to these airline disruption problems.

The multiple fleet airline disruption problems consist of two subproblems. In the first subproblem, sets of feasible flight routes for each fleet are generated by Dang and Ye's method [28] in a distributed computation network. In the second subproblem the feasible flight routes are reassigned to the available aircrafts in each fleet to form a recovery plan. In this paper, we mainly focus on the first subproblem and propose a modified traveling salesman problem (TSP) model to formulate the first subproblem. The flight legs are matched to form flight pairs, which are the decision variables in the modified TSP model, and this model is different from any model formulated for planning or recovery purposes in the literature. The solution of the problem is a sequence of flight pairs, and any two consecutive flight pairs are joint in the same flight legs. A distributed computation is proposed based on Dang and Ye's iterative method for integer programming. Using this method, the feasible flight routes for an aircraft are obtained sequentially in the lexicographical order from the original flight route of the aircraft. In the distributed computation the feasible flight routes for all the aircrafts in all the fleet types can be generated simultaneously. Not all feasible flight routes are needed for the second subproblem. Only by providing partial feasible flight routes can a solution of the second subproblem be found. When sufficient feasible flight routes are generated, solutions that are better than those in the literature can be found from the computational experience in Section 5.3.

The rest of this chapter is organized as follows. In Section 5.1, we give the mathematical formulations of the two subproblems. Dang and Ye's method and the distributed computation are introduced in Section 5.2. In Section 5.3, we present performance comparisons. Finally, a conclusion of this chapter is given in Section 5.4.

5.1 Problem formulation

The problem we focus on in this chapter is to reassign the available aircrafts during a disruption period. The disruption is caused by severe weather or aircraft mechanical malfunction or any other reasons and results in that some aircrafts are unavailable for a certain amount of time in their original schedule or even the whole day. All the flights served by the unavailable aircrafts may be canceled if no recovery action is taken. The cancellations can cost the air carriers a lot and dissatisfy the passengers. However, through reassigning the flight legs of the unavailable aircrafts to the remaining available aircrafts, a recovery schedule of the airline with a minimum impact caused by the disruption can be found.

Normally, most air carriers have multiple fleet types, and they have different characteristics. Of all the characteristics, the passenger capacity is the most important one to this research. During the recovery process of the disrupted airline schedule, an affected flight leg can only be reassigned to an aircraft with a passenger capacity that is larger than the actual passenger load of the flight leg.

The existence of aircraft substitutions in the multiple-fleet airline disruption problem makes it more difficult than the single-fleet problem since the sets of the flight legs that can be assigned to each fleet are different. To solve this problem, a distributed implementation of the iterative method [28] for integer programming is proposed to generate feasible flight routes for each fleet. Then the generated feasible flight routes are used to construct an aircraft reassignment to recovery the disrupted schedule. To find an aircrafts reassignment that minimizes the loss of the disruption and the deviation from the original schedule, the process concerns two subproblems as illustrated in Fig. 5.1. First, sets of the feasible flight routes for each fleet are generated simultaneously in the first subproblem formulated in Subsection 5.1.1. Then for the aircrafts in each fleet, feasible flight routes are chosen from the feasible flight route set that corresponds to the fleet to form a recovery schedule in the second subproblem given in Subsection 5.1.2.

5.1.1 Feasible flight routes generation

The decision variables in the feasible flight routes generation model of our previous works [29–31] are flight legs, and the sequence of flight legs cannot be decided in the previous models. Thus when the numerical solutions of the model are transformed into the actual flight routes, plenty of the actual flight routes are useless because there exists very long delay or grounding

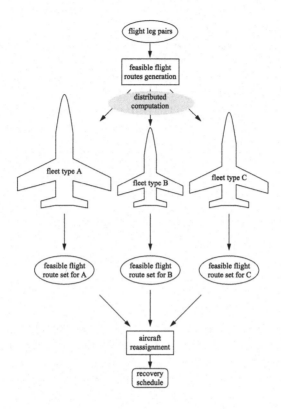

Figure 5.1 Process of two subproblems.

time between two flight legs. Thus enormous feasible solutions are needed to generate the actual feasible flight routes. To improve the efficiency, in this chapter, we propose a modified TSP model to generate feasible flight routes. The original flight legs are matched to form flight pairs, which are the decision variables in the modified TSP model. The flight pairs are generated under the constraint that the destination of the first flight leg is the same as the origin of the second flight leg. A sample schedule [32] in Table 5.1 is used to illustrate the generation of the flight pairs.

The TSP requires that all the cities have to be visited exactly once, and the visit returns to the beginning city to form a tour. However, all the flight legs are chosen to form a route at most once in the feasible flight routes generation problem. So some of the flight legs may not be chosen. We introduce two dummy stations, Source and Sink, to indicate the first and last flight legs of a flight route, respectively. Source combines with the original station of each flight route, and Sink combines with the

Table 5.1 Sample schedule.

Aircraft	Flight ID	Origin	Desti-nation	Depar-ture time	Arrival time	Dura-tion	Cancel-lation cost
	11	BOI	SEA	1410	1520	1:10	7350
A	12	SEA	GEG	1605	1700	0:55	10 231
	13	GEG	SEA	1740	1840	1:00	7434
	14	SEA	BOI	1920	2035	1:15	14 191
	21	SEA	BOI	1545	1700	1:15	11 189
B	22	BOI	SEA	1740	1850	1:10	12 985
	23	SEA	GEG	1930	2030	1:00	11 491
	24	GEG	SEA	2115	2215	1:00	9581
	31	GEG	PDX	1515	1620	1:05	9996
C	32	PDX	GEG	1730	1830	1:00	15 180
	33	GEG	PDX	1910	2020	1:10	17 375
	34	PDX	GEG	2100	2155	0:55	15 624

terminal station of each flight route in the original schedule to form dummy flight legs. The dummy flight legs match with the flight legs in the original schedule to form dummy flight pairs as in Table 5.2. Each resulting feasible flight pair route must originate from one of the source flight pairs and terminate at one of the sink flight pairs.

Table 5.3 displays the flight pairs and the dummy flight pairs constructed by the flight and dummy flight legs. The symbol $\sqrt{}$ in Table 5.3 is a valid flight pair, which consists of two consecutive flight legs whereas the symbol $\times$ is an invalid flight pair. The flight legs in the second column are the first flight legs in the flight pairs, and the flight legs in the second row are the second flight legs in the flight pairs. If an aircraft has flown the first flight leg in a flight pair, then it must fly the second flight leg in the flight pair. Some of the second flight legs in the flight pairs are delayed for a long period of time, such as the flight pair consisting of flight legs 34 and 13, and some of the turnaround time between the two flight legs in the flight pairs are too long, such as the flight pair consisting of flight legs 11 and 23. They can be marked $\times$ if the maximum delay time and the maximum turnaround time are given.

Fig. 5.2 is the network of the feasible flight route generation problem. The arcs represent the valid flight pairs, the valid dummy source flight pairs, and the valid dummy sink flight pairs, which are marked by $\sqrt{}$ in Table 5.3. For the arcs with a single arrow, only one flying direction is allowed. For the

Table 5.2 Dummy flight pairs.

Source flight pair		Sink flight pair	
First	Second	First	Second
Source→B ID:0	B→S 11	S→B 14	B→Sink ID:3
	B→S 22	S→B 21	
Source→S ID:1	S→G 12	B→S 11	S→Sink ID:4
	S→B 14	G→S 13	
	S→B 21	B→S 22	
	S→G 23	G→S 24	
Source→G ID:2	G→S 13	S→G 12	G→Sink ID:5
	G→S 24	S→G 23	
	G→P 31	P→G 32	
	G→P 33	P→G 34	

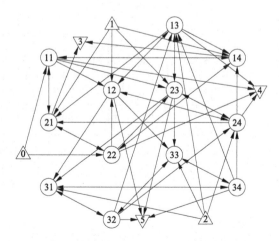

Figure 5.2 Network of feasible flight route generation problem.

Table 5.3 Full flight pair matrix.

		Flight ID												Dummy sink flight ID		
		11	12	13	14	21	22	23	24	31	32	33	34	3	4	5
Flight ID	11	×	✓	×	✓	✓	×	✓	×	×	×	×	×	×	✓	×
	12	×	×	✓	×	×	×	✓	✓	✓	×	✓	×	×	×	✓
	13	×	✓	×	✓	✓	×	✓	×	×	×	×	×	×	×	×
	14	✓	×	×	×	×	✓	×	×	×	×	×	×	✓	✓	×
	21	✓	×	×	×	×	✓	×	×	×	×	×	×	✓	×	×
	22	×	✓	×	✓	✓	×	✓	×	×	×	×	×	×	✓	×
	23	×	×	✓	×	×	×	✓	✓	✓	×	✓	×	×	×	✓
	24	×	✓	×	✓	✓	×	✓	×	×	×	×	×	×	✓	×
	31	×	×	✓	×	×	×	×	✓	✓	✓	✓	✓	×	×	×
	32	×	×	×	×	×	×	×	×	×	×	×	×	×	×	✓
	33	×	×	×	×	×	×	×	✓	✓	✓	✓	✓	×	×	×
	34	×	×	✓	×	×	✓	×	×	✓	×	✓	×	×	×	✓
Dummy source flight ID	0	✓	×	×	×	×	×	×	×	×	×	×	×	×	×	×
	1	×	✓	×	✓	✓	×	✓	×	×	×	×	×	×	×	×
	2	×	×	✓	×	×	×	×	✓	✓	×	✓	×	×	×	×

arcs with double arrows, both of the two flying directions are allowed, and they represent two valid flight pairs. The circular nodes represent the flight legs, and the triangular nodes and the inverse triangular nodes represent, respectively, the dummy source flight legs and the dummy sink flight legs. The problem is to find a set of feasible flight route for an involved fleet type, and each feasible flight route is a combination of the arcs in Fig. 5.2. Only partial arcs and nodes are chosen to form the combination, and they must satisfy the following conditions:

- Only nodes with actual passenger loads nonexceeding the aircraft capacity of the involved fleet can be chosen;
- Only one of the arcs that leave the triangular nodes is chosen;
- Only one of the arcs that enter the inverse triangular nodes is chosen;
- Each circular node in the chosen arcs is entered and left exactly once;
- No subtour exists.

Once all the requirements are met, the resulting chosen arcs are connected to each other in a sequence that originates from one of the triangular nodes and terminates at one of the inverse triangular nodes. The actual passenger loads of each node are within the aircraft capacity of the fleet type, so the feasible flight routes obtained from the chosen arcs can be flown by fleet-type aircrafts.

To mathematically formulate the network in Fig. 5.2, we denote by $F = \{1, \ldots, i, \ldots, j, \ldots, n\}$ the set of circular nodes that represent the flight legs and by $\mathscr{A}$ the set of arcs that represent the valid flight pairs $((i, j) \in \mathscr{A})$. We introduce the decision variables x_{ij} to formulate the feasible generation problem, and they denote the valid flight pairs: $x_{ij} = 1$ if the flight leg j immediately follows the flight leg i in the flight route, and $x_{ij} = 0$ otherwise. For the invalid flight pairs $(i, j) \notin \mathscr{A}$, which are marked by $\times$ in Table 5.3, we always have $x_{ij} = 0$.

Let $F_{\text{source}} = \{1, \ldots, u, \ldots, l\}$ denote the set of triangular nodes that represent the dummy source flight legs, let $\mathscr{A}_{\text{source}}$ denote the set of arcs that represent the valid dummy source flight pairs $((u, i) \in \mathscr{A}_{\text{source}})$, let $F_{\text{sink}} = \{1, \ldots, v, \ldots, m\}$ denote the set of inverse triangular nodes that represent the dummy sink flight legs, and let $\mathscr{A}_{\text{sink}}$ denote the set of arcs that represent the valid dummy sink flight pairs $((i, v) \in \mathscr{A}_{\text{sink}})$. The decision variables s_{ui} and k_{iv} denote the valid dummy source flight pairs and the valid dummy sink flight pairs, respectively: $s_{ui} = 1$ if the first flight leg of a flight route is i, and $s_{ui} = 0$ otherwise; $k_{iv} = 1$ if the last flight leg of a flight route is i, and $k_{iv} = 0$ otherwise. For the invalid dummy source flight pairs $((u, i) \notin \mathscr{A}_{\text{source}})$ and the invalid dummy sink flight pairs $((i, v) \notin \mathscr{A}_{\text{sink}})$, which are marked by $\times$ in Table 5.3, we always have $s_{ui} = 0$ and $k_{iv} = 0$.

The notations used in the formulation are as follows.

Indices

i, j, w	flight leg indices
u	dummy source flight index
v	dummy sink flight index

Sets

F	set of the flight legs
F_{source}	set of the dummy source flight legs
F_{sink}	set of the dummy sink flight legs
$\mathscr{A}$	set of the flight pairs
$\mathscr{A}_{\text{source}}$	set of the dummy source flight pairs
$\mathscr{A}_{\text{sink}}$	set of the dummy sink flight pairs

Coefficients

$f_{\min}$	minimum number of flight legs required by the problem
$f_{\max}$	maximum number of flight legs required by the problem
t_i	duration of the flight leg i including the turnaround time
T	total length of time from the departure of the earliest flight to the departure curfew time of stations
C	aircraft capacity of the involved fleet
c_i	actual passenger loads of the flight leg i

Variables

x_{ij}	equal to 1 if the flight leg j immediately follows the flight leg i in the flight route; 0 otherwise
s_{ui}	equal to 1 if the first flight leg of the flight route is i; 0 otherwise
k_{iv}	equal to 1 if the last flight leg of the flight route is i; 0 otherwise

Feasible route generation mathematical formulation

$$\exists x_{ij}, s_{ui}, k_{iv} \in \{0, 1\}$$

$$\forall (i, j) \in \mathscr{A}, (u, i) \in \mathscr{A}_{\text{source}}, (i, v) \in \mathscr{A}_{\text{sink}} \tag{5.1a}$$

subject to

(flight number constraint)

$$f_{\min} - 1 \le \sum_{(i,j) \in \mathscr{A}} x_{ij} \le f_{\max} - 1, \tag{5.1b}$$

(flight time constraint)

$$\sum_{(i,j)\in\mathscr{A}} t_j x_{ij} + \sum_{(u,i)\in\mathscr{A}_{\text{source}}} t_i s_{ui} \leq T, \tag{5.1c}$$

(sink node conservation)

$$\sum_{(i:(i,j)\in\mathscr{A})} x_{ij} - k_{jv} \geq 0 \ \forall(j,v) \in \mathscr{A}_{\text{sink}}, \tag{5.1d}$$

(source node conservation)

$$\sum_{(j:(i,j)\in\mathscr{A})} x_{ij} - s_{ui} \geq 0 \ \forall(u,i) \in \mathscr{A}_{\text{source}}, \tag{5.1e}$$

(flow-out node conservation)

$$\sum_{(w:(j,w)\in\mathscr{A})} x_{jw} + \sum_{(v:(j,v)\in\mathscr{A}_{\text{sink}})} k_{jv} - x_{ij} \geq 0 \ \forall(i,j) \in \mathscr{A}, \tag{5.1f}$$

(flow-in node conservation)

$$\sum_{(w:(w,i)\in\mathscr{A})} x_{wi} + \sum_{(u:(u,i)\in\mathscr{A}_{\text{source}})} s_{ui} - x_{ij} \geq 0 \ \forall(i,j) \in \mathscr{A}, \tag{5.1g}$$

(at most one flow-in node)

$$\sum_{(j:(i,j)\in\mathscr{A})} x_{ij} + \sum_{(v:(i,v)\in\mathscr{A}_{\text{sink}})} k_{iv} \leq 1 \ \forall i \in F, \tag{5.1h}$$

(at most one flow-out node)

$$\sum_{(i:(i,j)\in\mathscr{A})} x_{ij} + \sum_{(u:(u,j)\in\mathscr{A}_{\text{source}})} s_{uj} \leq 1 \ \forall j \in F, \tag{5.1i}$$

(source node cover)

$$\sum_{(u,i)\in\mathscr{A}_{\text{source}}} s_{ui} = 1, \tag{5.1j}$$

(sink node cover)

$$\sum_{(i,v)\in\mathscr{A}_{\text{sink}}} k_{iv} = 1, \tag{5.1k}$$

(subtour elimination)

$$\sum_{((i,j)\in\mathscr{A}:i\in U,j\in U)} x_{ij} \leq |U| - 1 \ \forall U \subset F \text{ such that } 2 \leq |U| \leq |F| - 2, \tag{5.1l}$$

(capacity constraint)

$$\sum_{((i,j)\in\mathscr{A}:c_i>C||c_j>C)} x_{ij} = 0. \tag{5.1m}$$

Prior to the construction of this model, the sequence of the flight pairs is rearranged according to the position of the disrupted flight legs. If some earlier flight legs in a flight route are disrupted, then the flight pairs are rearranged in the decreasing order of the departure time of the first flight leg. If the departure times of the first two flight legs of flight pairs are the same, then the departure times of the second flight legs are compared. For an original flight route, new feasible flight routes with different early flight legs can be found using Dang and Ye's method. If the disrupted flight legs appear in the back of a flight route, then the flight pairs are rearranged in the increasing order of the departure time of the first flight leg. Dang and Ye's method can compute new feasible flight routes with different back flight legs when an original flight route is given. If the whole flight route is disrupted for a large problem, then both the increasing and decreasing orders of the departure time sequence participate in the computation.

The model aims at enumerating all the feasible flight pair routes under constraints (5.1b)–(5.1l). Constraint (5.1b) limits the number of the flight legs to the requirement given by the problem. Constraint (5.1c) is a knapsack problem constraint, which ensures that the total durations of each chosen flight leg do not exceed the total length of time from the departure of the earliest flight leg to the departure curfew time of stations. Constraints (5.1d) and (5.1e) ensure that if a dummy sink or source flight pair is chosen, then there must be a flight pair connecting it. Constraint (5.1f) ensures that if a flight pair is chosen, then there must be a flight pair or a dummy sink flight pair connecting it on the second flight leg of the chosen flight pair. Constraint (5.1g) ensures that if a flight pair is chosen, then there must be a flight pair or a dummy source flight pair connecting it on the first flight leg of the chosen flight pair. Constraints (5.1h) and (5.1i) restrict each flight leg to be flown at most once. Constraints (5.1j) and (5.1k) impose that only one source node and one sink node exist. Constraint (5.1l) [33] is used to eliminate subtours. Constraint (5.1m) excludes flight pairs that consist of flight legs with actual passenger loads exceeding the aircraft capacity of the involved fleet. The dummy flight pairs consisting of flight legs with actual passenger loads exceeding the aircraft capacity can be excluded under constraints (5.1d), (5.1e), and (5.1m).

Constraints (5.1h) and (5.1i) are modified from the TSP constraint, which ensures that each city is entered and left exactly once [33]. They meet the requirement that each flight leg is flown at most once, but they cannot ensure that each flight pair is connected by another flight pair. For the first/second flight leg in a chosen flight pair, maybe none of the flight pairs that contain it as the second/first flight leg is chosen. So constraints (5.1d)–(5.1g) are constructed to ensure that all the chosen flight pairs connect to each other only once, resulting in that each flight leg in the chosen flight pairs is flown exactly once.

In [33] the authors claim that the number of constraint (5.1l) is nearly $2^{|F|}$. Hiwever, the number is limited to a small amount in the airline disruption problem. Let $U_s = \{(i,j) \in \mathscr{A} : i \in U, j \in U\}$, so that $|U_s| = |U|$ in the TSP. However, we always have $|U_s| \leq |U|$ in the airline disruption problem. Consider the situation where $U = \{i, j\}$ and $|U| = 2$. If x_{ij} is a valid flight pair $((i,j) \in \mathscr{A})$, then x_{ji} can be an invalid flight pair $((j, i) \notin \mathscr{A})$. The destination of flight leg i is the same as the origin of flight leg j, but the destination of flight leg j may be not the same as the origin of flight leg i. Besides, the change of the flight leg sequence from x_{ij} to x_{ji} can result in a huge delay time between two flight legs in x_{ji}. Once the delay time exceeds the maximum delay time given by the problem, x_{ji} should be classified as an invalid flight pair. Both $x_{11,14}$ and $x_{14,11}$ are valid flight pairs in Table 5.3, but the flight leg 11 has to delay nearly 4.5 hours in the flight pair $x_{14,11}$, so the latter should be classified as an invalid flight pair once a maximum delay time is given and the maximum delay time is shorter than the delay time in $x_{14,11}$. Since x_{ji} does not exist, we always have $|U_s| = 1 \leq |U| - 1$, and constraint (5.1l), which does not work, can be deleted for this situation.

The feasible solutions of (5.1) are sequences of the flight pairs that begin in a dummy source flight pair and end in a dummy sink flight pair, and they can be easily transformed to sequences of the flight legs that are the feasible flight routes. Although constraint (5.1c) ensures that the total duration of each chosen flight leg does not exceed the total length of available flying time, the actual total flying time of a feasible flight route is much longer than the sum of the durations of each flight leg in the feasible flight route. The reason is that there may exist big time intervals between two consecutive flight legs in the feasible flight route. Generally speaking, an aircraft cannot depart earlier than the original departure time in the recovery schedule, and it has to wait in the airport during the time intervals. If the total flying time of a feasible flight route is too long that the departure time of the last flight leg exceeds the departure curfew time of stations, then the feasible flight route will be deleted.

Let P be the set of all the feasible flight routes including the original flight routes, and let S be the set of all the airports. The research [34] asserts that $|P|$ is very large with respect to $|F|$ and $|S|$. If the length of each flight route is restricted to at most v flight legs, where $v \ll |F|$ and $v < |S|$, then

[34] shows that $|P|$ is bounded below by the function

$$\Phi(v) = O(2^v). \tag{5.2}$$

Thus $|P|$ is exponential with respect to the maximum flight route length v. No detail on generating P is given in [15,34,35]. In this paper, we propose the feasibility problem (5.1) to generate P. For small $|F|$ and $|S|$, it is not difficult and time-consuming to generate all the feasible flight routes using CPLEX CP Optimizer. However, when $|F|$ and $|S|$ are large, $|P|$ is extremely large, and CPLEX CP Optimizer cannot generate all the feasible flight routes in a reasonable amount of time. To tackle this difficulty, we propose a distributed computational approach to integer programming based on Dang and Ye's method to solve the feasibility problem (5.1). Details of this method are given in Section 5.2.

5.1.2 Aircrafts reassignment

Once the first subproblem (5.1) is solved, the feasible flight routes for one involved aircraft fleet can be obtained. Therefore the first subproblem has to be solved as many times as the number of the fleet types to obtain all the feasible flight route sets for each fleet type. Using the distributed computation introduced in Section 5.2, the first subproblem for each fleet type can be solved simultaneously. So all the feasible flight route sets for each fleet type can be obtained only once. With the feasible flight route sets, the second subproblem can be modeled as a resource assignment problem similar to the model in [34,35]. However, these papers reveal that the resource assignment problem is extremely difficult to solve since it is impossible to enumerate all the feasible flight routes. In this paper, we generate the feasible flight routes by Dang and Ye's method and order them in the lexicographical order with respect to the original flight routes. Thus by providing some partial feasible flight routes of problem (5.1) we can easily solve the aircraft reassignment problem using CPLEX Optimizers Concert Technology. There is no need to compute all the feasible flight routes. Details of the formulation for the aircrafts reassignment problem are given below.

We use the following notations in the formulation.

Indices

i	flight leg index
r	feasible flight route index
e	aircraft fleet type index
o	station index

Sets

F set of the flight legs

P^e set of the feasible flight routes for the fleet type e

Q set of the aircraft fleet types

S set of the stations

Coefficients

b_{ir} 1 if the flight leg i is included in the feasible flight route r; 0 otherwise

h_{or} 1 if the feasible flight route r departs from the station o; 0 otherwise

g_{or} 1 if the feasible flight route r terminates at the station o; 0 otherwise

d_r^e cost of reassigning the feasible flight route r to the fleet type e

q_i cost of canceling the flight leg i

H_o number of the aircrafts required to departure from the source stations o

G_o number of the aircrafts required to terminate at the sink stations o

A^e number of the available aircrafts in the fleet type e

Variables

y_r^e 1 if the feasible flight route r is reassigned to the fleet type e; 0 otherwise

z_i 1 if the flight i is canceled; 0 otherwise

Aircrafts reassignment mathematical formulation

$$\text{minimize} \sum_{e \in Q} \sum_{r \in P^e} d_r^e y_r^e + \sum_{i \in F} q_i z_i \tag{5.3a}$$

subject to

(flight cover)

$$\sum_{e \in Q} \sum_{r \in P^e} b_{ir} y_r^e + z_i = 1 \ \forall i \in F, \tag{5.3b}$$

(aircrafts balance of source stations)

$$\sum_{e \in Q} \sum_{r \in P^e} h_{or} y_r^e = H_o \ \forall o \in S, \tag{5.3c}$$

(aircrafts balance of sink stations)

$$\sum_{e \in Q} \sum_{r \in P^e} g_{or} y_r^e = G_o \ \forall o \in S, \tag{5.3d}$$

(resource utilization)

$$\sum_{r \in P^e} \gamma_r^e = A^e \ \forall e \in Q, \tag{5.3e}$$

(binary aircraft assignment)

$$\gamma_r^e \in \{0, 1\} \ \forall r \in P^e, e \in Q, \tag{5.3f}$$

(binary cancelation assignment)

$$z_i \in \{0, 1\} \ \forall i \in F. \tag{5.3g}$$

The coefficients b_{ir}, h_{or}, and g_{or} can be determined from the feasible solutions of Subsection 5.1.1. The remaining coefficients are inputs derived from the flight schedule.

The objective of (5.3a) is to minimize the total cost of each chosen flight route and total cancellation cost of each canceled flight leg. The flight cover constraint (5.3b) ensures that all the flight legs must be either in a flight route or canceled. The aircrafts balance of source stations constraint (5.3c) and the aircrafts balance of sink stations constraint (5.3d) implement the requirement for the number of the aircrafts positioned at the beginning and the end of the disruption period for each station. Constraint (5.3e) ensures that the number of the chosen feasible flight routes for each fleet type is equal to the number of the available aircrafts in the fleet type. Constraints (5.3f) and (5.3g) preclude fractional solutions.

5.2 Methodology

Problem (5.1) in Subsection 5.1.1 can also be solved by CPLEX as problem (5.3) in Subsection 5.1.2, but for a large airline disruption problem, it is unlikely to obtain all the feasible flight routes of problem (5.1) by CPLEX CP Optimizer in a reasonable amount of time as mentioned in [34,35]. The experiments from our former works [30,31,36] also show that it is impossible to enumerate all the solutions of problem (5.1) in a reasonable amount of time.

To tackle this difficulty, the initial seed cluster division method is introduced to divide the solution space of problem (5.1) into several segments, and each divided segment has a unique initial seed. Then Dang and Ye's method is applied to enumerate the feasible solutions of problem (5.1) in each divided segment. The obtained feasible solutions of problem (5.1) in a segment are only related to the initial seed of the segment, so the computation process of Dang and Ye's method in each divided segment is independent of each other. We propose a distributed computation such that the computation in each divided segment can be conducted simultaneously.

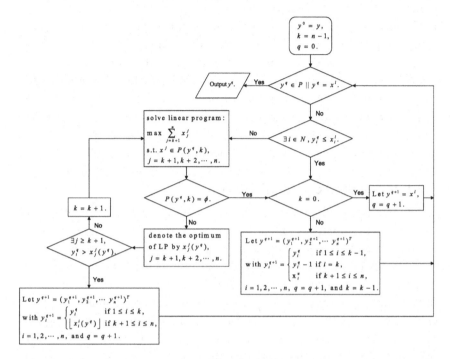

Figure 5.3 Flow diagram of the iterative method

Let $P = \{x \in R^n | Ax + Gw \leq b$ for some $w \in R^p\}$, where $A \in R^{m \times n}$ is an $m \times n$ integer matrix with $n \geq 2$, $G \in R^{m \times p}$ is an $m \times p$ matrix, and b is a vector of R^m.

Let $x^{\max} = (x_1^{\max}, x_2^{\max}, \ldots, x_n^{\max})^T$ with $x_j^{\max} = \max_{x \in P} x_j$, $j = 1, 2, \ldots, n$, and $x^{\min} = (x_1^{\min}, x_2^{\min}, \ldots, x_n^{\min})^T$ with $x_j^{\min} = \min_{x \in P} x_j$, $j = 1, 2, \ldots, n$.

Let $D(P) = \{x \in Z^n | x^l \leq x \leq x^u\}$, where $x^u = \lfloor x^{\max} \rfloor$ and $x^l = \lfloor x^{\min} \rfloor$.

For $z \in R^n$ and $k \in N_0$, let $P(z, k) = \{x \in P | x_i = z_i, 1 \leq i \leq k,$ and $x_i \leq z_i, k+1 \leq i \leq n\}$.

Given an integer point $y \in D(P)$ with $y_1 > x_i^l$, we present the iterative method from [28] in Fig. 5.3. It determines whether there is an integer point $x^* \in P$ such that $x^* \leq_l y$.

Dang and Ye's method is the improvement of the Dang's method [37]. The results obtained by both two methods are the same, but the computation processes are different. The idea of Dang and Ye's method to solve the integer programming is to define an increasing mapping from the lattice into itself. The integer points outside P are mapped into the first point in P that is smaller than them in the lexicographical order or x^l. All the integer points inside the polytope are fixed points under this increasing mapping. Given an initial integer point, the method either yields an integer point in the polytope or proves that no such point exists within a finite number of iterations. With a simple modification, all the integer points in P can be obtained sequentially in the lexicographical order. That is to say, for two consecutively obtained points, the one obtained earlier is always larger than that obtained later in the lexicographical order, and there is no other feasible point exists between two consecutively obtained points. For more details and proofs about this iterative method, we refer to [28,37].

5.2.1 Initial seed cluster division

Given another polytope P in Fig. 5.4, a lattice $D(P)$ can be easily constructed by the points x^u and x^l as illustrated in Fig. 5.4. Taking x^u as y^0 in Fig. 5.3, all the integer points in the polytope P can be enumerated in the lexicographical order using Dang and Ye's method. Providing several initial seed points as the big dots in Fig. 5.4 and taking each of them as y^0 in Fig. 5.3, respectively, the feasible integer points that are smaller than the seed points in the lexicographical order can be obtained using Dang and Ye's method. With a simple modification of Dang and Ye's method, the feasible integer points that are larger than the seed point in the lexicographical

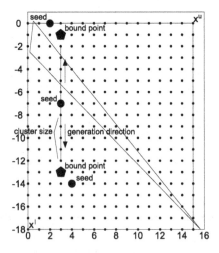

Figure 5.4 Initial seed cluster division.

order can also be obtained using Dang and Ye's method. So starting from the seed point, the enumeration for the feasible points can be conducted in two directions as in Fig. 5.4. The obtained feasible points form a point cluster around a seed point.

Let $x^{b_i} = (x_1^{b_i}, x_2^{b_i}, \ldots, x_n^{b_i})$, $i = 0, 1, \ldots, m$, denote bound points as the pentagon in Fig. 5.4, and let $S(P)_i = \{x \in Z^n | x^{b_{i-1}} <_l x \leq_l x^{b_i}\}$, $i = 1, 2, \ldots, m$, denote the segments defined by two consecutive bound points. Each segment $S(P)_i$ must satisfy the following two conditions:

$$D(P) = \cup_{i=1}^m S(P)_i \cup x^l, \tag{5.4a}$$

$$O = S(P)_i \cap S(P)_j \quad \text{for } i, j = 1, 2, \ldots, m, \; i \neq j. \tag{5.4b}$$

If the bound points x^{b_i} are calculated, then all the segments $S(P)_i$ are obtained and can be used to enumerate the feasible points in a distributed computation. Let $x^{b_0} = x^l$ and $x^{b_m} = x^u$. The calculation process of the remaining bound points is given below.

Let x^{l_i} and $x^{l_{i+1}}$ with $x^l <_l x^{l_i} <_l x^{l_{i+1}} <_l x^u$ be two consecutive initial seeds. The bound point x^{b_i} is in the middle of the x^{l_i} and $x^{l_{i+1}}$ in the lexicographical order. Thus $x^l <_l x^{l_i} <_l x^{b_i} <_l x^{l_{i+1}} <_l x^u$. Let $B(P)_i = \{x \in Z^n | x^{b_i} \leq_l x <_l x^u\}$, $I(P)_i = \{x \in Z^n | x^{l_i} \leq_l x <_l x^u\}$, and $s^{b_i} = |B(P)_i| = \left\lfloor \frac{|I(P)_i| + |I(P)_{i+1}|}{2} \right\rfloor$.
The bound points x^{b_i}, $i = 1, 2, \ldots, m-1$, are given by Lemma 6.

Lemma 6. *Let* $x^u = (x_1^u, x_2^u, \ldots, x_j^u, \ldots, x_n^u)$, $x^l = (x_1^l, x_2^l, \ldots, x_j^l, \ldots, x_n^l)$, $p_j = x_j^u - x_j^l + 1$ *for* $j = 1, 2, \ldots, n$, *and let* $s^{b_i} = |B(P)_i|$ *for* $i = 1, 2, \ldots, m-1$. *Assume that* $x^l \leq x^u$. *Then* $x^{b_i} = (x_1^{b_i}, x_2^{b_i}, \ldots, x_j^{b_i}, \ldots, x_n^{b_i})$ *for* $i = 1, 2, \ldots, m-1$ *in* $D(P)$ *are given by the formula*

$$x_j^{b_i} = \begin{cases} x_j^u - \left\lfloor \dfrac{s^{b_i} \bmod \prod\limits_{k=j}^n p_k}{\prod\limits_{k=j+1}^n p_k} \right\rfloor & \text{for } j = 1, 2, \ldots, n-1, \\ x_n^u - s^{b_i} \bmod p_n & \text{for } j = n. \end{cases} \tag{5.5}$$

For the proof of Lemma 6, see our previous work [30]. Once the bound points are calculated, all the segments are decided. The generation of the feasible points starts from the seed point in each segment using Dang and Ye's method, and the generation goes in both the direction that is lexicographically larger than the seed and the direction that is lexicographically smaller than the seed, as illustrated in Fig. 5.4. The parameter cluster size is used to control the number of the feasible points to be obtained in one direction in each segment, so the generation in one of the two directions stops when the number of the obtained feasible points equals the cluster size or a bound point is encountered. The set of the obtained feasible points in each segment forms a cluster around the seed point in the segment, and the cluster is a subset of the segment.

This division method is especially devised to solve the feasibility problem (5.1) in Subsection 5.1.1. When a disruption happens, some of the original flight routes may become irregular and result in losses of revenue. Alternative flight routes of the irregular original flight routes can minimize the losses. So the feasibility problem (5.1) is formulated as finding the alternative flight routes for one fleet type. Take the original flight routes as the initial seed points in Fig. 5.4. The polytope P and lattice $D(P)$ can be constructed from the constraints of problem (5.1), and the computation of problem (5.1) for one fleet type can be transformed into the generation of the feasible points in the polytope P in Fig. 5.4 using the method from [28]. Normally, there are more than one aircraft in a fleet type, so $D(P)$ can be divided into several segments, and the generation of the feasible flight routes from the original flight route in each segment can be conducted simultaneously using the method from [28] in a distributed computation network introduced in Subsection 5.2.2.

Dang and Ye's method generates new feasible points, which are always lexicographically smaller than the one generated earlier, and it can also generate new feasible points that are always lexicographically larger than the one generated earlier if the generation direction is reversed. So both the feasible flight routes that are lexicographically larger than the original flight routes and the feasible flight routes that are lexicographically smaller than the original flight routes can be found. Each segment $S(P)_i$ can be divided again by another division method proposed in [30] if more computation processors are available.

5.2.2 Distributed computation implementation

A distributed computation network is constructed by the message passing interface (MPI) and consists of plenty of computers. MPI is used to construct the communication among the processors of each computer. Due to enough computation processors, the distributed computation can increase the computation efficiency and it can be conducted in three different ways:
- The feasibility problem (5.1) for each fleet type,
- The generation of the feasible flight routes in each divided segment for one feasibility problem (5.1),

- The linear programming in Dang and Ye's method as illustrated in Fig. 5.3.

The feasible flight routes are generated only for one fleet type when the feasibility problem (5.1) is solved once. So problem (5.1) has to be solved as many times as the number of the fleet types to obtain all the feasible flight route sets for all the fleet types. Problem (5.1) for each fleet type differs in the matrices of polytope P that are constructed from constraints (5.1b)–(5.1m) and the original flight routes. After the matrices and the original flight routes of each fleet type are sent to each participated computers in the computation network, the distributed computation can be started. The communication of data can be easily done by the send operation MPI_SEND and the receive operation MPI_RECV from MPI, and the control of all the participated computers can also be easily done by MPI. Details of the usage of MPI can be found in "A Message-Passing Interface Standard" [38].

The $D(P)$ of each feasibility problem (5.1) can be divided into several segments, and these segments satisfy condition (5.4). Besides, the obtained feasible flight routes of problem (5.1) in a segment is only related to the original flight route of the segment. So the computation process of Dang and Ye's method in each divided segment is independent of each other and can be conducted simultaneously in each computation processor. Details of the distributed computation among segments of one feasibility problem (5.1) can be found in [30].

The linear programming in Dang and Ye's method aims at maximizing the sum of a series of variables. Since these variables are independent to each other, they can be maximized individually, and the results are the same. Thus the linear programming in Fig. 5.3 can be reformulated as follows:

$$\text{maximize} \quad x_j^j \qquad \text{for } j = k+1, k+2, \ldots, n, \qquad (5.6a)$$

$$\text{subject to} \quad x^j \in P(\gamma^q, k). \qquad (5.6b)$$

The number of the linear programming increases from 1 to $n-k$, which can be solved simultaneously in the distributed computation network if there are enough computation processors. It can also be easily done by MPI, and the implementations are conducted on a simple problem that has few variables. No implementation on the problems in Section 5.3 is conducted due to the limited computation processors. The distributed computation of the linear programming can indeed save a lot of computation time for large-degree problems with enough computation processors provided.

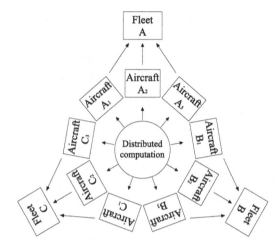

Figure 5.5 Implementation of distributed computation.

In an implementation the previous two ways of the distributed computations can be treated together, as illustrated in Fig. 5.5. Since each segment is divided based on an original flight route of an aircraft. The total segments for all the aircrafts in all the fleet types can be obtained first. The generation of the feasible flight routes starts from the original flight routes within the bound points in each segment. So for the generation of the feasible flight routes for the aircrafts in one fleet type, the differences are the bound points that define the segments and the original flight routes of the aircrafts. For the generation of the feasible flight routes for all the aircrafts in all the fleet types, the differences are the bound points that define the segments, the original flight routes of the aircrafts, and the matrices of polytope Ps. The distributed computation is implemented by sending the bound points, the original flight routes, and the matrices of polytope P to each computation processors in the distributed computation network.

5.3 Computational experience

Airline flight schedules that appear in [16,39] are used to evaluate our approach to the airline disruption problems, and details of the schedules can be found in [39] and [16]. The distributed computation network constructed by MPI consisting of 3 different computers; one computer can establish 16 threads simultaneously to compute, and the other two can establish 2 threads. So at most 20 segments can be processed simultaneously. All the programs are coded in C++, and CPLEX Concert Technology is used to solve the linear programming by Dang and Ye's method. The version of the Cplex used is 12.6.1. First, problem (5.1) in Subsection 5.1.1 is solved using the method of [28]. Then all these feasible flight routes are used to generate a solution of problem (5.3) in Subsection 5.1.2 by CPLEX Optimizers's Concert Technology.

In our previous works [30,31], numerical results have shown that both Dang's method and Dang and Ye's method outperform CPLEX CP Optimizer when they are used to solve problem (5.1). So no comparison is made between Dang and Ye's method and CPLEX CP Optimizer in this chapter. We mainly focus on the comparison of the present results with those in [39] and [16].

5.3.1 European midsize airline results

Paper [39] employs a European midsize airline schedule to demonstrate their approach. The schedule consists of 29 aircrafts from 10 different fleet types: e Airbus 319, 320, 321, 322, and 32C, Boeing 736, 738, 739, and 73G, and Fokker F70. Each of the fleet has a different passenger capacity. 126 flight legs are served by all the aircrafts between 43 stations. The lattice $D(P)$ of problem (5.1) is divided into 29 segments, and each segment has a unique seed point, which is the original flight route and a unique pair of bound points. For the segments whose corresponding aircrafts are in the same fleet type, they share the same matrix of polytope P. Since only 20 segments can be processed simultaneously, the last 9 segments can be started to compute after anyone of the first 20 segments is finished.

A disruption given in research [39] happens when the aircraft 321-YYLBC is breakdown and under repair from 8:00 AM until 11:00 PM. So the disruption lasts for 15 hours. The flight legs XX863 VIE-CAI, XX864 CAI-VIE, and XX127 VIE-FRA, which are originally assigned to the aircraft 321-YYLBC, are affected by the disruption. These flight legs may be canceled if no recovery plan is made. The cancellation of the flight leg XX127 VIE-FRA causes aircrafts unbalance at the airport FRA at the end of the schedule. So a recovery plan is needed to make sure that there is an aircraft substituting the aircraft 321-YYLBC and terminating at the airport FRA at the end of the schedule. Besides, the cancellations of the flight legs result in huge losses of revenue, so the recovery plan also aims at minimizing the losses.

Tables 5.4 and 5.5 are the solutions obtained by our approach and the comparison with the solutions given in [39]. Row two to row four of

Table 5.4 Comparison with [39].

Method	Cluster size	Delayed flights	Total delay minutes	Flight swaps	Unaltered flight routes
Wu	2	0	0	9	24
	5	0	0	8	24
	10	0	0	5	26
Babic	\	2	255	3	26
		3	361	5	25
		2	483	5	25
Dispatch-ers' solution	\	0	0	5	26

Table 5.5 Details of solutions obtained by our approach.

Method	Cluster size	Details of solutions
Wu	2	321-YYLBA: XX863, XX864, XX127 738-YYLNR: XX1425, XX9741, XX9742, XX1426, XX9767, XX9768 738-YYLNS: XX9863, XX9864, XX125, XX126 320-YYLBT: XX458, XX801, XX802, XX417, XX418
	5	321-YYLBA: XX863, XX864, XX127 738-YYLNR: XX1425, XX9741, XX9742, XX1426, XX307 320-YYLBO: XX308, XX125, XX126 320-YYLBT: XX458, XX801, XX802, XX417, XX418
	10	321-YYLBA: XX863, XX864, XX127 738-YYLNR: XX1425, XX9741, XX9742, XX1426, XX417, XX418

Table 5.4 are the solutions of problem (5.3) obtained by our approach, and each solution is a recovery plan of the airline disruption problem. The solutions are computed by the feasible flight routes generated in problem (5.1.1). The parameter cluster size in the second column is used to control the number of the feasible flight routes generated in both directions in each segment. So the total number of the feasible flight routes obtained for each available aircraft equals the double of the cluster size. Column three and column four are the number of the delayed flight legs and the total delay

time in minutes, respectively. Column five is the number of the flight legs swapped with other flight legs of flight routes. Column six is the number of the flight routes not affected by the disruption, and they are flown as the plans in the original schedule.

The cluster size is enlarged to test the impact of the number of the feasible flight routes on the solution of problem (5.3). A larger cluster size can find a better solution of the disruption problem, because more feasible flight routes involved in the constitution of the recovery plan means more possibilities that a better recovery plan with less flight swaps and more unaltered flight routes can be produced. When the cluster size increases to 10, the solution, which is the same as the dispatchers' solution in row eight, is found. The recovery plan obtained under the cluster size 10 is what the airline dispatchers would implement.

Row five to row seven of Table 5.4 are solutions computed in [39]. There are delayed flight legs in each of the recovery plan they produced. Although there are less flight swaps in their solutions, the amount of the total delayed flight time brings a huge losses of revenue to airline. The airline dispatchers would not implement any of these solutions computed in [39]. The dispatchers' solution that would be implemented by the dispatchers in row eight is given not through computation. It is obtained by modifying the solution in the row seven of Table 5.4.

Details of solutions obtained by our approach can be found in Table 5.5. The aircrafts appeared in Table 5.5 fly the flight legs that are reassigned to them, and the remaining aircrafts fly their original flight route in the recovery plan. The solutions without delayed flight legs can be found only under the cluster size 2, which means that only 4 feasible flight routes are needed for each available aircraft to produce the recovery plan. 20 feasible flight routes are needed for each available aircraft to produce the recovery plan that the airline dispatchers would implement. Also, all the solutions computed in [39] in row five to row seven of Table 5.4 can be constituted using the feasible flight routes obtained under the cluster size 10. This is done by setting the parameter "SolnPoolAGap" of CPLEX Optimizer Concert Technology to a proper value. A larger value of the parameter "SolnPoolAGap" means that more nonoptimal solutions can be obtained. Also, nonoptimal solutions that deviate too much from the optimum can be found by setting a larger "SolnPoolAGap".

Our approach is more effective than the method of [39] when solving this European midsize airline schedule disruption problem: the dispatchers'

Table 5.6 Swedish domestic airline schedule characteristics.

	s1	s2
Number of aircraft	13	30
Number of aircraft types	2	5
Number of flight legs	98	215
Number of airports	19	32

Table 5.7 Causes of the disruptions.

Disruption problems	Causes of the disruptions
$s1a$	An unavailability of aircraft for 5 hours
$s1b$	Imposed delays on two flight legs for 25 and 30 minutes
$s2a$	An unavailability of aircraft for 6 hours
$s2b$	Imposed delays on four flight legs for 15, 20, 30, and 40 minutes

solution can be computed by our approach, but it cannot be computed by the approach proposed in [39].

5.3.2 Swedish domestic airline results

The research [16] uses two flight schedules, $s1$ and $s2$, from Swedish domestic airline to illustrate their approach. The characteristics of the two schedules are given in Table 5.6. The two schedules are disrupted in two different ways, a and b, given in [16]; a is disrupted by an unavailability of an aircraft, and b is disrupted by an imposed delay on several flight legs. So there are four disruption problems $s1a$, $s1b$, $s2a$, and $s2b$. The causes of the disruptions are given in Table 5.7.

The paper [16] aims at maximizing the revenue of airline when solving the disruption problems, using weights to calculate the costs of the disruptions. So the objective function (5.3a) of problem (5.3) is modified to maximize the revenue of the airline.

We use the following notations in the formulation of the modified objective function.

Weights

can	cancellation weight
sw_t	weight for swaps between the flight legs in different fleet types
sw	weight for swaps between the flight legs in the same fleet types
del	delay weight

Coefficients

st_r number of swaps between the flight legs in different fleet types in the feasible flight route r

s_r number of swaps between the flight legs in the same fleet types in the feasible flight route r

d_r delay in passenger minutes in the feasible flight route r

With the weights and new coefficients, the modified objective function is

$$\text{maximize} \sum_{e \in Q} \sum_{r \in P^e} (can \times \sum_{i \in F} b_{ir} c_i - sw\_t \times \mathbf{st_r} - sw \times \mathbf{s_r} - del \times \mathbf{d_r}) y_r^e. \quad (5.7)$$

So the aircraft reassignment problem in Subsection 5.1.2 is to maximize objective (5.7) under constraints (5.3b)–(5.3g) for these disruptions problems from the Swedish domestic airline. To be convenient for comparison, the notations of the weights and new coefficients are the same as those in [16]. To distinguish them from the notations used before, the new coefficients are in bold. For example, $\mathbf{d_r}$ is different from d_r^e, which appears in problem (5.3) in Subsection 5.1.2.

Table 5.8 Sets of weights.

Problem	Weights			
	can	sw_t	sw	del
s1a1	20	100	10	1
s1a2	20	1000	10	1
s1a3	100	1000	10	1
s1a4	100	100	10	1
s1b1	20	100	10	1
s1b2	20	100	100	1
s1b3	20	100	400	1
s2a1	20	100	10	1
s2a2	20	1000	10	1
s2a3	100	1000	10	1
s2b1	20	100	10	1
s2b2	20	100	10	10
s2b3	20	100	50	1

Objective (5.7) is the sum of the revenues of the feasible flight routes. The revenue of each feasible flight route is calculated by subtracting the swap cost $sw\_t \times \mathbf{st_r}$, $sw \times \mathbf{s_r}$ and delay cost $del \times \mathbf{d_r}$ from the cancellation weight can multiplied by the actual passenger load $\sum_{i \in F} b_{ir} c_i$ on the feasible flight route. The coefficients st_r and s_r can be inspected from each feasible flight route. For a delayed flight leg, the delay in passenger minutes equals the actual passenger load on the delayed flight leg multiplied by the delay time of the delayed flight leg. So the coefficient d_r is calculated by summing the delays in passenger minutes of each delayed flight leg in the feasible flight route r.

The paper [16] provides several different weight settings for each of the four disruption problems $s1a$, $s1b$, $s2a$, and $s2b$ as in Table 5.8. Thus more problem instances can be generated from the original four disruption problems. $s1a1$ uses flight schedule $s1$ with disruption a and weight setting 1, which sets can to 20, sw_t to 100, sw to 20, and del to 1 to generate a disruption problem. A unique solution with different characteristics can be obtained with each weight setting, and all of the 13 generated problems are used to evaluate our approach.

Table 5.9 Results of $s1$.

Problem	Cluster size	Obj	Characteristics			
			c	st	s	d
$s1a1$	3	42 660	32	0	0	1420
	10	43 760	46	0	4	0
	50	43 800	25	4	2	0
	2200	43 840	30	2	8	0
$s1a2$	3	42 660	32	0	0	1420
	10	43 760	46	0	4	0
$s1a3$	3	218 980	32	0	0	1420
$s1a4$	3	218 980	32	0	0	1420
	50	220 680	25	4	2	0
$s1b1$	5	43 140	0	0	0	1580
	25	43 340	0	0	5	1330
	75	43 620	0	0	15	950
	100	44 445	0	0	10	175
	125	44 650	0	0	7	0
$s1b2$	5	43 140	0	0	0	1580
	100	43 870	0	0	6	250
	125	44 070	0	0	4	250
$s1b3$	5	43 140	0	0	0	1580

Tables 5.9 and 5.10 are the results of the problems generated from the schedules $s1$ and $s2$, respectively. Obj is the solution of the objective function (5.7) under constraints (5.3b)–(5.3g), c is the number of the canceled

Table 5.10 Results of s2.

Problem	Cluster size	Obj	Characteristics			
			c	st	s	d
s2a1	260	68 480	29	2	4	0
s2a2	5	68 000	65	0	0	0
s2a3	1500	341 865	29	0	12	1615
s2b1	1500	69 080	0	0	12	100
s2b2	1500	68 720	23	0	12	0
s2b3	*10*	*68 950*	*0*	*2*	*1*	*100*

passengers in the solution, *st* is the number of swaps between the flight legs in different fleet types in the solution, *s* is the number of swaps between the flight legs in the same fleet types in the solution, and *d* is the delay in passenger minutes in the solution.

In Table 5.9 the solutions of each problem generated from schedule s1 under different cluster sizes are given. The cluster size is increased to find a better solution of each problem since more feasible flight routes can constitute a better solution. Once a solution is equal to or better than that computed by [16], the cluster size stops increasing. The row in bold is the best solution of each problem, and a solution, which is better than that in [16], is found for the problem s1b1 in the row marked in bold and italic. It is obtained under the cluster size 125, and the objective value is larger than that [16] by 50. The details of the solution are offered below. Five aircrafts fly the flight legs reassigned to them, and the remaining aircrafts that did not appear below fly their original flight routes in the recovery plan.

- SELEA: 22, 23, 77, 78, 24, 25, 26, 27, 28, 29, 30
- SELEC: 42, 43, 44, 52, 53
- SELED: 46, 47, 48, 49, 50, 51, 45
- SELEZ: 76, 89, 90, 79, 80, 81, 82
- SELIP: 88, 91, 92, 93, 94, 95, 96, 97

In Table 5.10 the final solution of each problem generated from schedule s2 is given. The first five solutions equal those computed by [16]. The solution of the last problem s2b3 is better than that of [16] by 50, and it is computed only under the cluster size 10. Details of the solution of the problem s2b3 are given below.

- SEISY: 37, 40, 41, 42, 43, 44,45, 46
- SELED: 120, 121, 122, 123, 124, 125, 126, 127
- SELIN: 200, 201, 202, 203, 204, 205, 206, 128
- SELIP: 207, 38, 39, 208, 209, 210, 211, 212, 213, 214

We can see that present approach is effective to solve the airline disruption problems. Not only the solutions that are equal to the solutions in the literature can be computed, but also the solutions that are better than the solutions in the literature can be found for some problems. It is because each feasible flight route is generated from an original flight route, and the feasible flight routes can be obtained sequentially in the lexicographical order using Dang and Ye's method. The feasible flight routes have relations with the original flight routes. Some of the flight legs in the feasible flight routes are different from those in the original flight routes, whereas others are the same. The number of different flight legs in the feasible flight routes increases as the generation for the feasible flight routes goes on. Sometimes, the feasible flight routes with more flight legs that are different from those in the original flight routes can constitute a better solution of the disruption problem, and more feasible flight routes have to be generated before the feasible flight routes with more flight legs different from those in the original flight routes can be obtained. So the quality of the solutions are related to the number of the feasible flight routes controlled by the cluster size. Some problems only need a few feasible flight routes to obtain a better solution of the disruption problem, but some other problems may need much more. Basically speaking, a much better solution of the disruption problem can be obtained by providing more feasible flight routes. In an actual implementation the cluster size can be set to a relatively larger value, and the computation of the disruption problem can be conducted periodically. For example, the aircraft reassignment problem (5.3) in Subsection 5.1.2 can be solved once 100 new feasible flight routes are obtained. The computation stops when the solution satisfies the airline dispatchers.

5.4 Conclusion

A novel modified TSP model is formulated to generate feasible flight routes for producing a recovery plan for the airline disruption problems. A distributed computation network is constructed to solve the modified TSP model using Dang and Ye's method, and the feasible flight routes for all the aircrafts in all the fleet types can be generated simultaneously in the distributed computation. Dang and Ye's method ensures that the obtained feasible flight routes are in the lexicographical order automatically, and there does not exist other feasible flight route between two consecutively obtained feasible flight routes. This characteristic enables the generated feasible flight routes to constitute a recover plan easily, and only partial feasible flight routes are needed. If more feasible flight routes are provided, then a much better recover plan can be constituted. In the evaluation of present approach, not only the same solutions as those in the literature can be computed, but also better solutions can be found. More details about the method can be found in [40].

References

[1] Jens Clausen, Allan Larsen, Jesper Larsen, Natalia J. Rezanova, Disruption management in the airline industry – concepts, models and methods, Computers & Operations Research 37 (May 2010) 809–821.

[2] Ahmed Abdelghany, Khaled Abdelghany, Modeling Applications in the Airline Industry, Ashgate, 2010.

[3] Sheng-Chen Huang, Airline schedule recovery following disturbances: an organizationally-oriented decision-making approach, Ph.D. dissertation, University of California, Berkeley, Berkeley, CA, 2005.

[4] Michael D. Clarke, Barry C. Smith, Impact of operations research on the evolution of the airline industry, Journal of Aircraft 41 (1) (2004) 62–72.

[5] Civil Aviation Administration of China, Statistical communiqué on civil aviation industry development in 2010, Air Transport & Business 292 (2011) 19–24.

[6] Jinfu Zhu, Air Transportation Operation, Northwestern Polytechnical University Press, Xi'an, China, 2009.

[7] CAAC News, Liu Shaoyong: the airline operating cost increases by cny1,000 for each minute' delay, Civil Aviation Administration of China News, http://editor.caacnews.com.cn/mhb/html/2012-03/28/content_93332.htm, 2012.

[8] EUROCONTROL, Performance review report: an assessment of air traffic management in Europe during the calendar year 2011, EUROCONTROL. Performance Review Commission, Brussels, http://www.eurocontrol.int/sites/default/files/content/documents/single-sky/pru/publications/prr/prr-2011-draft.pdf, 2012.

[9] CODA, Delays to air transport in Europe, CODA. Digest Annual 2011, Brussels, http://www.eurocontrol.int/documents/coda-digest-annual-2011, 2012.

[10] Air Transport Association. Annual and per-minute cost of delays to U.S. airlines, http://www.airlines.org/Pages/Annual-and-Per-Minute-Cost-of-Delays-to-U.S.-Airlines.aspx, 2011.

[11] Alexander Niehues, Sören Belin, Tom Hansson, Richard Hauser, Mercedes Mostajo, Julia Richter, Punctuality: How Airlines Can Improve on-Time Performance, Booz Allen & Hamilton Inc., McLean, VA., USA, 2001.

[12] Dušan Teodorović, Slobodan Guberinić, Optimal dispatching strategy on an airline network after a schedule perturbation, European Journal of Operational Research 15 (2) (1984) 178–182.

[13] Dušan Teodorović, Goran Stojković, Model for operational daily airline scheduling, Transportation Planning and Technology 14 (4) (1990) 273–285.

[14] Dušan Teodorovič, Goran Stojkovič, Model to reduce airline schedule disturbances, Journal of Transportation Engineering 121 (4) (1995) 324–331.

[15] Jay M. Rosenberger, Ellis L. Johnson, George L. Nemhauser, Rerouting aircraft for airline recovery, Transportation Science 37 (2003) 408–421.

[16] Tobias Andersson, Peter Värbrand, The flight perturbation problem, Transportation Planning and Technology 27 (2) (2004) 91–117.

[17] Shangyao Yan, Chinhui Tang, Chonglan Shieh, A simulation framework for evaluating airline temporary schedule adjustments following incidents, Transportation Planning and Technology 28 (3) (2005) 189–211.

[18] Michael Løve, Kim Riis Sørensen, Jesper Larsen, Jens Clausen, Disruption management for an airline – rescheduling of aircraft, in: Stefano Cagnoni, Jens Gottlieb,
Emma Hart, Martin Middendorf, Günther Raidl (Eds.), Applications of Evolutionary
Computing, in: Lecture Notes in Computer Science, vol. 2279, Springer, Berlin/Hei-
delberg, 2002, pp. 289–300.
[19] Michael F. Arguello, Jonathan F. Bard, Gang Yu, Models and methods for managing
airline irregular operations, in: Gang Yu (Ed.), Operations Research in the Airline Industry, in: International Series in Operations Research & Management Science, vol. 9,
Springer, 1998, pp. 1–45.
[20] Jonathan Bard, Gang Yu, Michael Argüello, Optimizing aircraft routings in response
to groundings and delays, IIE Transactions 33 (2001) 931–947.
[21] Ladislav Lettovský, Airline operations recovery: an optimization approach, Ph.D. dissertation, Department of Industrial and Systems Engineering, Georgia Institute of
Technology, Atlanta, GA, 1997.
[22] Niklas Kohl, Allan Larsen, Jesper Larsen, Alex Ross, Sergey Tiourine, Airline disruption management – perspectives, experiences and outlook, Journal of Air Transport
Management 13 (3) (2007) 149–162.
[23] Tungkuan Liu, Chiruey Jeng, Yuhern Chang, Disruption management of an
inequality-based multi-fleet airline schedule by a multi-objective genetic algorithm,
Transportation Planning and Technology 31 (6) (2008) 613–639.
[24] Bernard Munyazikwiye, Hamid Reza Karimi, Kjell Robbersmyr, Optimization of
vehicle-to-vehicle frontal crash model based on measured data using genetic algorithm, IEEE Access 5 (99) (2017) 3131–3138.
[25] Yanling Wei, Jianbin Qiu, Hamid Reza Karimi, Reliable output feedback control of
discrete-time fuzzy affine systems with actuator faults, IEEE Transactions on Circuits
and Systems 1 (2017) 170–181.
[26] Wenqiang Ji, Anqing Wang, Jianbin Qiu, Decentralized fixed-order piecewise affine
dynamic output feedback controller design for discrete-time nonlinear large-scale systems, IEEE Access 5 (99) (2017) 1977–1989.
[27] Wen Xing, Yuxin Zhao, Hamid Karimi, Convergence analysis on multi-AUV systems
with leader-follower architecture, IEEE Access 5 (99) (2017) 853–868.
[28] Chuangyin Dang, Yinyu Ye, A fixed point iterative approach to integer programming
and its distributed computation, Fixed Point Theory and Applications 2015 (1) (2015)
1–15.
[29] Benchi Li, Chuangyin Dang, Jingjing Zheng, Solving the large airline disruption
problems using a distributed computation approach to integer programming, in: International Conference on Information Science and Technology, 2013, pp. 444–450.
[30] Zhengtian Wu, Benchi Li, Chuangyin Dang, Fuyuan Hu, Qinxin Zhu, Baochuan
Fu, Solving long haul airline disruption problem caused by groundings using a distributed fixed-point computational approach to integer programming, Neurocomputing (2017).
[31] Zhengtian Wu, Benchi Li, Chuangyin Dang, Solving the large airline disruption
problems caused by the airports closure using a distributed-computation approach to
integer programming, Working paper, 2017.
[32] Benjamin G. Thengvall, Models and Solution Techniques for the Aircraft Schedule
Recovery Problem, Ph.d. Dissertation, Graduate Program in Operations Research
and Industrial Engineering, University of Texas, Austin, 1999.
[33] Laurence A. Wolsey, George L. Nemhauser, Integer and Combinatorial Optimization, Wiley-Interscience Series in Discrete Mathematics and Optimization., A Wiley-
Interscience Publication, 1988.
[34] Michael F. Argüello, Framework for exact solutions and heuristics for approximate
solutions to airlines' irregular operations control aircraft routing problem, Ph.D. dissertation, Department of Mechanical Engineering, University of Texas, Austin, 1997.

[35] Michael F. Argüello, Jonathan F. Bard, Gang Yu, A grasp for aircraft routing in response to groundings and delays, Journal of Combinatorial Optimization 1 (1997) 211–228.

[36] Zhengtian Wu, Chuangyin Dang, Hamid Reza Karimi, Changan Zhu, Qing Gao, A mixed 0–1 linear programming approach to the computation of all pure-strategy Nash equilibria of a finite n-person game in normal form, Mathematical Problems in Engineering 2014 (5) (2014) 1–8.

[37] Chuangyin Dang, An increasing-mapping approach to integer programming based on lexicographic ordering and linear programming, in: The Ninth International Symposium on Operations Research and Its Applications, (ISORA'10) Chengdu-Jiuzhaigou, China, August 19–23, 2010, pp. 55–60.

[38] Message Passing Interface Forum, MPI: a message-passing interface standard, version 2.2, http://www.mpi-forum.org/docs/mpi-2.2/mpi22-report.pdf, 2009.

[39] Obrad Babic, Milica Kalic, Danica Babic, Slavica Dozic, The airline schedule optimization model: validation and sensitivity analysis, Procedia – Social and Behavioral Sciences 20 (2011) 1029–1040.

[40] Zhengtian Wu, Benchi Li, Chuangyin Dang, Solving multiple fleet airline disruption problems using a distributed-computation approach to integer programming, IEEE Access 5 (2017) 19116–19131.

CHAPTER 6

A deterministic annealing neural network algorithm for the minimum concave cost transportation problem

6.1 Introduction

Transportation network design, which gives potential to decrease the total costs, plays an important role in logistic systems of many firms. Due to its theoretical and practical importance, the transportation problem is the core problem in the transportation network design. The target of the problem is to reduce the overall shipping cost between customers and suppliers. It is shown in the literature that even minor improvements of the transportation problem always imply substantial overall savings in the firm. If the transportation cost and transported amount of the product are linearly related, then the transportation problem can be solved in polynomial time via linear programming algorithms. In reality, however, the transportation costs are always nonlinear and frequently concave, where the unit cost for transported products reduces as the product amount increases [1]. The minimum concave cost transportation problem proposed in this chapter is about minimizing a concave quadratic function subject to transportation constraints, which is a well-known NP hard problem. To compute an optimal point of the minimum concave cost transportation problem, a number of traditional iterative algorithms have been developed in the literature such as Erickson et al. [2], Fang and Tsao [3], Gallo and Sodini [4], Ahmed et al. [5], and Gupta and Pavel [6]. A detailed survey of algorithms for the minimum concave cost transportation problem can be found in Guisewite [7] and Dang et al. [8]. To our knowledge, the traditional algorithms, such as generalized reduced gradient algorithm [9] and barrier algorithm [10], still get the best application in industry. However, most of these studies are mainly focused on the theory of the problem analysis, or the algorithms developed are very hard to be implemented. It is still a challenge to solve the minimum concave cost transportation problem effectively and efficiently.

Integer Optimization and Its Computation in Emergency Management
https://doi.org/10.1016/B978-0-32-395203-3.00011-3
119

Recently, a number of different algorithms have been developed for optimization problems. Among them, neural network is one of the most efficient algorithms, originated by Hopfield and Tank [11] and has been applied to a variety of optimization problems. The details of the theory of neural networks can be found in [12]. However, most problems tackled in the existing neural networks are devoted to linear programming problems or convex quadratic programming problems. Liu and Dang [13] proposed a new recurrent neural network for linear 0–1 programming problems, Hu and Zhang [14] presented a neural network for convex quadratic programming problems, Hu and Wang [15] proposed a dual neural network for a special class of convex quadratic programming problems, and He et al. [16] developed a recurrent neural network based on the method of penalty functions to solve a bilevel linear programming problem, after inserting the inertial term into a first-order projection neural network. He et al. [17] presented an inertial projection neural network to compute variational inequalities, and Rutishauser et al. [18] developed a distributed neocortical-like neuronal networks to solve constraint-satisfaction problems. Xu et al. [19–21] proposed neural nets based on a hybrid of Lagrange and transformation approaches. A systematic investigation of such neural networks for combinatorial optimization problems can be found in [22–27]. Moreover, an elastic network combinatorial optimization method was proposed by Durbin and Willshaw [28], and other optimization neural networks can be found in the literature; see, for instance, [29–34]. To the best of our knowledge, however, there are few works on neural network for concave quadratic transportation problems, which partly motivated us for this study.

The main contributions of this chapter are listed as follows: i) A novel neural network algorithm is proposed to find the solution of the concave cost transportation problem. ii) The proposed algorithm integrates the annealing scheme and Lagrange-barrier function into neural network for concave quadratic transportation problems. It is shown that the proposed neural network is completely stable and converges to a stable equilibrium state. The mathematical proof and some tests will be given in this chapter. iii) The numerical simulations demonstrate that the proposed neural network can converge to global or near-global optimal solutions. The algorithm described in this chapter can be potentially used to solve real-world problems.

The rest of this chapter is organized as follows. We introduce the proposed neural network models and propose a neural network annealing algorithm in Section 6.2. In Section 6.3, we prove that the proposed neural

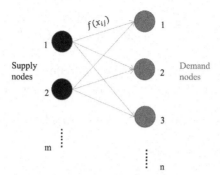

Figure 6.1 Transportation problem.

networks are completely stable and converge to a stable equilibrium state. To show that the proposed neural network annealing algorithm is effective and efficient, we conduct some numerical experiments in Section 6.4. We conclude the chapter with some remarks in Section 6.5.

6.2 Deterministic annealing and neural networks

As shown in Fig. 6.1, the transportation problem is described as a network. Let $(1, 2, \ldots, m)$ be the set of supply nodes, and let $(1, 2, \ldots, n)$ be for the set of demand nodes. Arc (i, j) stands for the arc from supply node i to demand j.

In this chapter, we mathematically state the minimum concave cost transportation problem as follows: Find a global minimum point of

Problem (I):

$$
\begin{aligned}
\min \quad & f(x) \\
\text{subject to} \quad & \sum_{j=1}^{n} x_{ij} = a_i, \ i = 1, \ldots, m, \\
& \sum_{i=1}^{m} x_{ij} = b_j, \ j = 1, \ldots, n, \\
& 0 \le x_{ij}, \ j = 1, \ldots, p_i, \\
& 0 \le x_{ij} \le u_{ij}, \ j = p_i, \ldots, n, \\
& i = 1, 2, \ldots, m,
\end{aligned}
\tag{6.1}
$$

where x_{ij} is the transporting amount of goods along arc (i, j), $f : R^{m \times n} \to R$ is a separable and continuously differentiable concave function, which stands for the cost function, $0 < a_i$, $0 < b_j$, u_{ij} is the flow upper bound for arc (i,j), and $\sum_{i=1}^{m} a_i = \sum_{j=1}^{n} b_j$, which implies that the sum of total supply is equal to that of the total demand. Assume that Problem (I) has a strictly feasible solution.

To solve (6.1), we incorporate the constraints $0 \leq x_{ij}$ and $0 \leq x_{ij} \leq u_{ij}$ into the objective function by introducing the barrier terms [20]

$$\beta x_{ij} \ln x_{ij} \text{ and } \beta \left(x_{ij} \ln x_{ij} + (u_{ij} - x_{ij}) \ln(u_{ij} - x_{ij})\right)$$

and hence obtain the following
   Problem (II):

$$\min \quad e(x, \beta)$$
$$\text{subject to} \quad \sum_{j=1}^{n} x_{ij} = a_i, i = 1, \ldots, m, \qquad (6.2)$$
$$\sum_{i=1}^{m} x_{ij} = b_j, j = 1, \ldots, n,$$

where

$$e(x, \beta) = f(x) + \beta \sum_{i=1}^{m} \left(\sum_{j=1}^{p_i} x_{ij} \ln x_{ij} \right.$$
$$\left. + \sum_{j=p_i+1}^{n} \left(x_{ij} \ln x_{ij} + (u_{ij} - x_{ij}) \ln(u_{ij} - x_{ij})\right)\right),$$

and β is a parameter, which varies from a sufficiently large positive number to zero.

Remark 1. When β approaches zero, Problem (II) is equivalent to Problem (I).

Consider the following Lagrange function of Problem (II):

$$L(x, \lambda) = f(x) + \sum_{i=1}^{m} \lambda_i^r \left(\sum_{j=1}^{n} x_{ij} - a_i \right)$$
$$+ \sum_{j=1}^{n} \lambda_j^c \left(\sum_{i=1}^{m} x_{ij} - b_j \right)$$
$$+ \beta \sum_{i=1}^{m} \left(\sum_{j=1}^{p_i} x_{ij} \ln x_{ij} \right. \qquad (6.3)$$
$$\left. + \sum_{j=p_i+1}^{n} \left(x_{ij} \ln x_{ij} + (u_{ij} - x_{ij}) \ln(u_{ij} - x_{ij})\right)\right),$$

where $\lambda = (\lambda_1^r, \ldots, \lambda_m^r, \lambda_1^c, \ldots, \lambda_n^c)^{\top}$ is the so-called Lagrange multiplier vector. Computing the partial derivative of $L(x, \lambda)$ with respect to x_{ij}, we obtain

$$\frac{\partial L(x, \lambda)}{\partial x_{ij}} = \frac{\partial f(x)}{\partial x_{ij}} + g_{ij}(x_{ij}, \lambda), \qquad (6.4)$$

where

$$g_{ij}(x_{ij}, \lambda) = \lambda_i^r + \lambda_j^c + \begin{cases} \beta(1 + \ln x_{ij}) & \text{if } 1 \leq j \leq p_i, \\ \beta \ln \frac{x_{ij}}{u_{ij} - x_{ij}} & \text{if } p_i + 1 \leq j \leq n. \end{cases}$$

Let

$$\nabla_x L(x, \lambda) = \left(\frac{\partial L(x, \lambda)}{\partial x_{11}}, \dots, \frac{\partial L(x, \lambda)}{\partial x_{mn}} \right)^\top,$$

which is the gradient of $L(x, \lambda)$ at x. The necessary optimality condition says that if x^* is an optimal solution of Problem (II), then there exists λ^* that together with x^* satisfies

$$\nabla_x L(x^*, \lambda^*) = 0,$$
$$\sum_{j=1}^n x_{ij}^* = a_i, \ i = 1, \dots, m, \tag{6.5}$$
$$\sum_{i=1}^m x_{ij}^* = b_j, \ j = 1, \dots, n.$$

A point x^* satisfying (6.5) is called a stationary point of (6.2).

For convenience, let

$$r_i = \exp\left(\frac{\lambda_i^r}{\beta} \right), \ c_j = \exp\left(\frac{\lambda_j^c}{\beta} \right),$$

$$r = (r_1, \dots, r_m)^\top, \ c = (c_1, \dots, c_n)^\top,$$

$$\alpha_{ij}(x) = \begin{cases} \exp(1 + \frac{1}{\beta} \frac{\partial f(x)}{\partial x_{ij}}) & \text{if } 1 \leq j \leq p_i, \\ \exp(\frac{1}{\beta} \frac{\partial f(x)}{\partial x_{ij}}) & \text{if } p_i + 1 \leq j \leq n, \end{cases}$$

for $i = 1, \dots, m$, $j = 1, \dots, n$. For any given $\beta > 0$, from $\frac{\partial L(x, \lambda)}{\partial x_{ij}} = 0$ we get

$$x_{ij} = \begin{cases} \frac{1}{r_i c_i \alpha_{ij}(x)} & \text{if } 1 \leq j \leq p_i, \\ \frac{u_{ij}}{1 + r_i c_i \alpha_{ij}(x)} & \text{if } p_i + 1 \leq j \leq n. \end{cases}$$

Let

$$d_{ij}(x) = \begin{cases} \frac{1}{r_i c_i \alpha_{ij}(x)} & \text{if } 1 \leq j \leq p_i, \\ \frac{u_{ij}}{1 + r_i c_i \alpha_{ij}(x)} & \text{if } p_i + 1 \leq j \leq n, \end{cases} \tag{6.6}$$

$i = 1, \dots, m$, $j = 1, \dots, n$, and

$$d(x) = \left(d_{11}(x), \dots, d_{1n}(x), \dots, d_{m1}(x), \dots, d_{mn}(x) \right)^\top.$$

Thus $\frac{\partial L(x,\lambda)}{\partial x_{ij}} = 0$ is equivalent to $d_{ij}(x) - x_{ij} = 0$, and $\nabla_x L(x, \lambda) = 0$ is equivalent to $d(x) - x = 0$.

To find a solution of Problem (II), we design a neural network model for the first equation of (6.5) as follows:

$$\frac{dx_{ij}}{dt} = d_{ij}(x) - x_{ij}, \quad i = 1, \ldots, m, j = 1, \ldots, n. \tag{6.7}$$

Remark 2. This neural network model plays an important role in the algorithm design. For one thing, this model can provide an iterative direction of x_{ij} in the proposed algorithm. For another, it is an indispensable part in the proof of convergence of the proposed algorithm.

Consider the function

$$E(x) = f(x) + \sum_{i,j} \int_0^{x_{ij}} g_{ij}(v, \lambda) dv. \tag{6.8}$$

If we can prove that $\frac{dE(x)}{dt} < 0$ for all x except for those equilibrium points ($\frac{dx}{dt} = 0$) where it vanishes, then $E(x)$ is a Lyapunov function for the neural network system (6.7), and the system will be completely stable and converges to a stable equilibrium state [35]. This will be realized as Theorem 1 in the next section.

Note that at an equilibrium point, i.e., at a point satisfying $\frac{dx}{dt} = 0$, we have $x = d(x)$, which is a solution of $\nabla_x L(x, \lambda) = 0$. Thus the minimization of the Lyapunov function $E(x)$ is equivalent to the minimization of $L(x, \lambda)$ with respect to x. Therefore an equilibrium point can be obtained by $\min_x L(x, \lambda)$, which can be realized by successive line search along the descent direction $d(x) - x$ (this will be proved in Theorem 1 in the next section).

Also, note that at the equilibrium point, λ (and therefore r and c) should satisfy the last two equations of (6.5). To get such r and c, we construct the following neural network model. Denote

$$x_{ij} = \omega(r_i, c_j) = \begin{cases} \frac{1}{r_i c_j \alpha_{ij}(x)} & \text{if } 1 \le j \le p_i, \\ \frac{u_{ij}}{1 + r_i c_j \alpha_{ij}(x)} & \text{if } p_i + 1 \le j \le n, \end{cases} \tag{6.9}$$

$i = 1, \ldots, m, j = 1, \ldots, n$. Substituting (6.9) into the second and third equations of formula (6.5) we obtain

$$
\begin{aligned}
\sum_{j=1}^{n} \omega(r_i, c_j) &= a_i, \quad i = 1, \ldots, m, \\
\sum_{i=1}^{m} \omega(r_i, c_j) &= b_j, \quad j = 1, \ldots, n.
\end{aligned}
\tag{6.10}
$$

Denote

$$
\begin{aligned}
u_i(r, c) &= r_i \left(\sum_{j=1}^{n} \omega(r_i, c_j) - a_i \right), \quad i = 1, \ldots, m, \\
v_j(r, c) &= c_j \left(\sum_{i=1}^{m} \omega(r_i, c_j) - b_j \right), \quad j = 1, \ldots, n, \\
u(r, c) &= \left(u_1(r, c), \ldots, u_m(r, c) \right)^{\mathsf{T}},
\end{aligned}
\tag{6.11}
$$

and

$$
v(r, c) = \left(v_1(r, c), \ldots, v_n(r, c) \right)^{\mathsf{T}}.
$$

Then we propose a neural network model as follows.

- State equation:

$$
\frac{d}{dt} \begin{pmatrix} r \\ c \end{pmatrix} = \begin{pmatrix} u(r, c) \\ v(r, c) \end{pmatrix}.
\tag{6.12}
$$

- Output equation:

$$
r = r, \quad c = c.
$$

The role of this neural network model is similar to that of (6.7). It can both provide the iterative direction of r, c in the proposed algorithm and play as an indispensable part in the proof of convergence of the proposed algorithm.

Consider the function

$$
E(r, c) = \frac{1}{2} \left(\sum_{i=1}^{m} \left(\sum_{j=1}^{n} \omega(r_i, c_j) - a_i \right)^2 + \sum_{j=1}^{n} \left(\sum_{i=1}^{m} \omega(r_i, c_j) - b_j \right)^2 \right).
\tag{6.13}
$$

Note that the solution of (6.10) is equivalent to the solution of $\min_{\{r, c\}} \{E(r, c)\}$. If we can prove that $E(r, c)$ is a Lyapunov function for the neural network system (6.12), then the neural network system will be completely stable and converge to a stable equilibrium state [35]. Therefore the solution of (6.10) (i.e., the solution of the last two equations of (6.5)) can be obtained by minimization of $E(r, c)$. It will also be proved in

Theorem 2 in the next section that $(u(r, c), v(r, c))$ is a descent direction of $E(r, c)$. To get an equilibrium point of the neural network (i.e., a solution of $\min_{\{r, c\}}\{E(r, c)\}$ or a solution of (6.10)), we propose the following iteration procedure.

Let

$$v = \min\left\{\frac{1}{a_1}, \ldots, \frac{1}{a_m}, \frac{1}{b_1}, \ldots, \frac{1}{b_n}\right\}.$$

Let (r^0, c^0) be an arbitrary positive vector, and for $k = 0, 1, \ldots$, and let

$$r_i^{k+1} = r_i^k + \mu_k r_i^k \left(\sum_{j=1}^{n} \omega(r_i^k, c_j^k) - a_i\right), \quad i = 1, \ldots, m,$$

$$c_j^{k+1} = c_j^k + \mu_k c_j^k \left(\sum_{i=1}^{m} \omega(r_i^k, c_j^k) - b_j\right), \quad j = 1, \ldots, n,$$

(6.14)

where $\mu_k \in (0, v]$ satisfy

$$E(r^{k+1}, c^{k+1}) < E(r^k, c^k).$$

It is obvious that when $\mu_k \in [0, v]$, r^{k+1} and c^{k+1} are always positive if $0 < r^k$ and $0 < c^k$.

Then (6.14) can be rewritten as

$$r^{k+1} = r^k + \mu_k u(r^k, c^k) \tag{6.15}$$

and

$$c^{k+1} = c^k + \mu_k v(r^k, c^k). \tag{6.16}$$

It will be proved in Theorem 3 in the next section that the above procedure will converge to the solution of $\min_{\{r, c\}}\{E(r, c)\}$ or the solution of (6.10). Based on the above-mentioned results, we can propose the following neural network annealing algorithm (briefly NNAA) for Problem (I).

Algorithm (NNAA).

Step 0: *Let β be a sufficiently large positive number such that $e(x, \beta)$ is strictly convex. Let x^0 be an arbitrary point satisfying $0 < x_{ij}^0$, $j = 1, \ldots, p_i$, and $0 < x_{ij}^0 < u_{ij}$, $j = p_i + 1, \ldots, n$, $i = 1, \ldots, m$. Take (r^0, c^0) to be an arbitrary positive vector. Given $x = x^0$, let $(r^*(x^0), c^*(x^0))$ be a positive approximate solution of neural network (6.12)–(6.13), which is obtained with (6.15)–(6.16) that starts at (r^0, c^0). Let $r^0 = r^*(x^0)$ and $c^0 = c^*(x^0)$, and compute*

$$x_{ij}^1 = \begin{cases} \frac{1}{r_i^*(x^0)c_j^*(x^0)\alpha_{ij}(x^0)} & \text{if } 1 \le j \le p_i, \\ \frac{u_{ij}}{1+r_i^*(x^0)c_j^*(x^0)\alpha_{ij}(x^0)} & \text{if } p_i + 1 \le j \le n, \end{cases} \tag{6.17}$$

$i = 1, \ldots, m, j = 1, \ldots, n$. Let $q = 1$ and go to Step 1.

Step 1: Given $x = x^q$, let $(r^*(x^q), c^*(x^q))$ be a positive approximate solution of neural network (6.12)–(6.13), which is obtained with (6.15)–(6.16) that starts at (r^0, c^0). Go to Step 2.

Step 2: Compute

$$d_{ij}(x^q) = \begin{cases} \frac{1}{r_i^*(x^q) c_i^*(x^q) \alpha_{ij}(x^q)} & \text{if } 1 \le j \le p_i, \\ \frac{u_{ij}}{1 + r_i^*(x^q) c_i^*(x^q) \alpha_{ij}(x^q)} & \text{if } p_i + 1 \le j \le n, \end{cases} \tag{6.18}$$

$i = 1, \ldots, m, j = 1, \ldots, n$. If $\|d(x^q) - x^q\|$ is sufficiently small, then terminate when β is small enough, or decrease β by a factor. Let $x^1 = x^q$, $r^0 = r^*(x^q)$, $c^0 = c^*(x^q)$, and $q = 1$, and go to Step 1. Otherwise, do as follows: Let

$$\lambda^*(x^q) = \left(\lambda_1^{r*}(x^q), \ldots, \lambda_m^{r*}(x^q), \lambda_1^{c*}(x^q), \ldots, \lambda_n^{c*}(x^q)\right)^\top \tag{6.19}$$

with $\lambda_i^{r*}(x^q) = \beta \ln r_i^*(x^q)$, $i = 1, \ldots, m$, and $\lambda_j^{c*}(x^q) = \beta \ln c_j^*(x^q)$, $j = 1, 2 \ldots, n$, and compute

$$x^{q+1} = x^q + \theta_q(d(x^q) - x^q), \tag{6.20}$$

where θ_q is a positive number in $(0, 1]$ satisfying

$$L(x^{q+1}, \lambda^*(x^q)) < L(x^q, \lambda^*(x^q)).$$

Let $q = q + 1$, and go to Step 1.

To present the proposed algorithm clearly, we provide a flowchart of this algorithm in Fig. 6.2.

In the next section, we will prove that the proposed neural network annealing algorithm converges to the solution of Problem (I).

6.3 Stability and convergence analysis of the proposed algorithm

In this section, we rigorously prove that the neural network defined in (6.7) and (6.8) is completely stable and converges to a stable equilibrium state. With this favorable property, the proposed algorithm generates a stable output when the initial conditions vary within a reasonable range, which has obvious significance in real applications.

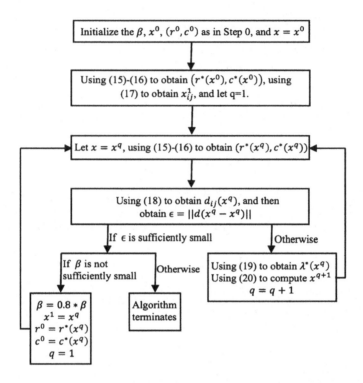

Figure 6.2 Flowchart of the NNAA algorithm.

Theorem 1. *Let* $E(x)$ *be a Lyapunov function for the network system satisfying* $\frac{dE(x)}{dt} < 0$ *for all* x *except the equilibrium points* $(\frac{dx}{dt} = 0)$. *Then the neural network system is completely stable and converges to a stable equilibrium state.*

Proof. From formulas (6.4), (6.7), and (6.8) we see that

$$
\begin{aligned}
\frac{dE(x)}{dt} &= \sum_{i,j} \frac{\partial f(x)}{\partial x_{ij}} \frac{dx_{ij}}{dt} + \sum_{i,j} g_{ij}(x_{ij}, \lambda) \frac{dx_{ij}}{dt} \\
&= \sum_{i,j} \left[\frac{\partial f(x)}{\partial x_{ij}} + g_{ij}(x_{ij}, \lambda) \right] \frac{dx_{ij}}{dt} \\
&= \sum_{i,j} \frac{\partial L(x, \lambda)}{\partial x_{ij}} \frac{dx_{ij}}{dt} \\
&= \sum_{i,j} \frac{\partial L(x, \lambda)}{\partial x_{ij}} \left[d_{ij}(x) - x_{ij} \right] \\
&= \nabla_x L(x, \lambda)^\top \left(d(x) - x \right).
\end{aligned} \tag{6.21}
$$

We will further prove that if $d_{ij}(x) - x_{ij} \neq 0$, then

$$\frac{\partial L(x, \lambda)}{\partial x_{ij}}\left(d_{ij}(x) - x_{ij}\right) < 0. \tag{6.22}$$

Thus, if $d(x) - x \neq 0$, then

$$\nabla_x L(x, \lambda)^\top \left(d(x) - x\right) < 0. \tag{6.23}$$

We can prove this in two cases:

1. Suppose $d_{ij}(x) - x_{ij} < 0$. Then

$$
\begin{aligned}
&1 < r_i c_j \alpha_{ij}(x) x_{ij} && \text{if } 1 \leq j \leq p_i, \\
&1 < r_i c_j \alpha_{ij}(x) \frac{x_{ij}}{u_{ij} - x_{ij}} && \text{if } p_i + 1 \leq j \leq n.
\end{aligned} \tag{6.24}
$$

Taking the natural logarithm of both sides of (6.24), we obtain

$$
\begin{aligned}
&0 < \frac{1}{\beta}\left(\frac{\partial f(x)}{\partial x_{ij}} + \lambda_i^r + \lambda_j^c\right) + 1 + \ln x_{ij} && \text{if } 1 \leq j \leq p_i, \\
&0 < \frac{1}{\beta}\left(\frac{\partial f(x)}{\partial x_{ij}} + \lambda_i^r + \lambda_j^c\right) + \ln \frac{x_{ij}}{u_{ij} - x_{ij}} && \text{if } p_i + 1 \leq j \leq n.
\end{aligned} \tag{6.25}
$$

Multiplying by $\beta > 0$ both sides of (6.25), we obtain

$$
\begin{aligned}
&0 < \frac{\partial f(x)}{\partial x_{ij}} + \lambda_i^r + \lambda_j^c + \beta(1 + \ln x_{ij}) = \frac{\partial L(x, \lambda)}{\partial x_{ij}} && \text{if } 1 \leq j \leq p_i, \\
&0 < \frac{\partial f(x)}{\partial x_{ij}} + \lambda_i^r + \lambda_j^c + \beta \ln \frac{x_{ij}}{u_{ij} - x_{ij}} = \frac{\partial L(x, \lambda)}{\partial x_{ij}} && \text{if } p_i + 1 \leq j \leq n.
\end{aligned}
$$

$$\tag{6.26}$$

Thus, when $d_{ij}(x) - x_{ij} < 0$,

$$\frac{\partial L(x, \lambda)}{\partial x_{ij}}\left(d_{ij}(x) - x_{ij}\right) < 0. \tag{6.27}$$

2. Suppose $d_{ij}(x) - x_{ij} > 0$. Then

$$
\begin{aligned}
&1 > r_i c_i \alpha_{ij}(x) x_{ij} && \text{if } 1 \leq j \leq p_i, \\
&1 > r_i c_i \alpha_{ij}(x) \frac{x_{ij}}{u_{ij} - x_{ij}} && \text{if } p_i + 1 \leq j \leq n.
\end{aligned} \tag{6.28}
$$

Taking the natural logarithm of both sides of (6.28), we obtain

$$0 > \frac{1}{\beta}\left(\frac{\partial f(x)}{\partial x_{ij}} + \lambda_i^r + \lambda_j^c\right) + 1 + \ln x_{ij} \qquad \text{if } 1 \leq j \leq p_i,$$

$$0 > \frac{1}{\beta}\left(\frac{\partial f(x)}{\partial x_{ij}} + \lambda_i^r + \lambda_j^c\right) + \ln \frac{x_{ij}}{u_{ij} - x_{ij}} \qquad \text{if } p_i + 1 \leq j \leq n.$$

$$(6.29)$$

Multiplying by $\beta > 0$ both sides of (6.29), we obtain

$$0 > \frac{\partial f(x)}{\partial x_{ij}} + \lambda_i^r + \lambda_j^c + \beta(1 + \ln x_{ij}) = \frac{\partial L(x, \lambda)}{\partial x_{ij}} \qquad \text{if } 1 \leq j \leq p_i,$$

$$0 > \frac{\partial f(x)}{\partial x_{ij}} + \lambda_i^r + \lambda_j^c + \beta \ln \frac{x_{ij}}{u_{ij} - x_{ij}} = \frac{\partial L(x, \lambda)}{\partial x_{ij}} \qquad \text{if } p_i + 1 \leq j \leq n.$$

$$(6.30)$$

Thus, when $d_{ij}(x) - x_{ij} > 0$,

$$\frac{\partial L(x, \lambda)}{\partial x_{ij}}\left(d_{ij}(x) - x_{ij}\right) < 0. \tag{6.31}$$

Observe that

$$\nabla_x L(x, \lambda)^\top \left(d(x) - x\right) = \sum_{i=1}^{m}\sum_{j=1}^{n} \frac{\partial L(x, \lambda)}{\partial x_{ij}}\left(d_{ij}(x) - x_{ij}\right) < 0.$$

Thus $\frac{dE(x)}{dt} < 0$. The last inequality also indicates that $d(x) - x$ is a descent direction of $L(x, \lambda)$. The conclusion is true in accordance with [35].   □

For the neural network (6.12)–(6.13), we will prove that it is completely stable and converges to a stable equilibrium state. Moreover, $(u(r, c)^\top, v(r, c)^\top)^\top$ is a descent direction of $E(r, c)$.

Theorem 2. $\frac{dE(r,c)}{dt} < 0$ for all (r, c) except the equilibrium points $(\frac{d}{dt}\binom{r}{c} = 0)$, and thus $E(r, c)$ is a Lyapunov function for the network system, and the neural network system is completely stable and converges to a stable equilibrium state.

Proof. Similarly to the proof of Theorem 1, we only need to prove that $\frac{dE(r,c)}{dt} < 0$ for all (r, c) except the equilibrium points. We can see from formula (6.12) that

$$\frac{dE(r, c)}{dt} = \nabla E(r, c)^\top \frac{d}{dt}\binom{r}{c}$$

$$= \nabla E(r, c)^\top \binom{u(r, c)}{v(r, c)}, \tag{6.32}$$

where

$$\nabla E(r, c) = \left(\frac{\partial E(r, c)}{\partial r_1}, \ldots, \frac{\partial E(r, c)}{\partial r_m}, \frac{\partial E(r, c)}{\partial c_1}, \ldots, \frac{\partial E(r, c)}{\partial c_n} \right)^{\top} \tag{6.33}$$

is the gradient of $E(r, c)$ at (r, c).

Computing the partial derivative of $E(r, c)$ with respect to r_l, we obtain

$$\begin{aligned}
\frac{\partial E(r, c)}{\partial r_l} &= \sum_{h=1}^{n} \left(\sum_{p=1}^{n} \omega(r_l, c_p) - a_l \right) \frac{\partial \omega(r_l, c_h)}{\partial r_l} \\
&+ \sum_{h=1}^{n} \left(\sum_{p=1}^{m} \omega(r_p, c_h) - b_h \right) \frac{\partial \omega(r_l, c_h)}{\partial r_l},
\end{aligned} \tag{6.34}$$

where

$$\frac{\partial \omega(r_l, c_h)}{\partial r_l} = \begin{cases} -\frac{c_h \alpha_{lh}(x)}{(\eta c_h \alpha_{lh}(x))^2} & \text{if } 1 \le h \le p_l, \\ -\frac{u_{lh} c_h \alpha_{lh}(x)}{(1 + \eta c_h \alpha_{lh}(x))^2} & \text{if } p_l + 1 \le h \le n. \end{cases} \tag{6.35}$$

Computing the partial derivative of $E(r, c)$ with respect to c_h, we obtain

$$\begin{aligned}
\frac{\partial E(r, c)}{\partial c_h} &= \sum_{l=1}^{m} \left(\sum_{p=1}^{n} \omega(r_l, c_p) - a_l \right) \frac{\partial \omega(r_l, c_h)}{\partial c_h} \\
&+ \sum_{l=1}^{m} \left(\sum_{p=1}^{m} \omega(r_p, c_h) - b_h \right) \frac{\partial \omega(r_l, c_h)}{\partial c_h},
\end{aligned} \tag{6.36}$$

where

$$\frac{\partial \omega(r_l, c_h)}{\partial c_h} = \begin{cases} -\frac{\eta \alpha_{lh}(x)}{(\eta c_h \alpha_{lh}(x))^2} & \text{if } 1 \le h \le p_l, \\ -\frac{u_{lh} \eta \alpha_{lh}(x)}{(1 + \eta c_h \alpha_{lh}(x))^2} & \text{if } p_l + 1 \le h \le n. \end{cases} \tag{6.37}$$

Then we have

$$\begin{aligned}
\nabla E(r, c)^{\top} & \begin{pmatrix} u(r, c) \\ v(r, c) \end{pmatrix} \\
= & \sum_{l=1}^{m} \sum_{h=1}^{n} \left(\sum_{p=1}^{n} \omega(r_1, c_p) - a_l \right)^2 r_l \frac{\partial \omega(r_l, c_h)}{\partial r_l} \\
& + \sum_{l=1}^{m} \sum_{h=1}^{n} \left(\sum_{p=1}^{n} \omega(r_1, c_p) - a_l \right) \left(\sum_{p=1}^{m} \omega(r_p, c_h) - b_h \right) r_l \frac{\partial \omega(r_l, c_h)}{\partial r_l} \\
& + \sum_{h=1}^{n} \sum_{l=1}^{m} \left(\sum_{p=1}^{m} \omega(r_p, c_h) - b_h \right) \left(\sum_{p=1}^{n} \omega(r_l, c_p) - a_l \right) c_h \frac{\partial \omega(r_l, c_h)}{\partial c_h} \\
& + \sum_{h=1}^{n} \sum_{l=1}^{m} \left(\sum_{p=1}^{m} \omega(r_p, c_h) - b_h \right)^2 c_h \frac{\partial \omega(r_l, c_h)}{\partial c_h}.
\end{aligned} \tag{6.38}$$

From (6.35) and (6.37) we obtain that

$$r_l \frac{\partial \omega(r_l, c_h)}{\partial r_l} = c_h \frac{\partial \omega(r_l, c_h)}{\partial c_h} < 0.$$

Observe that $\sum_h \sum_t t_{hl} = \sum_t \sum_h t_{hl}$. Thus

$$\nabla E(r, c)^\top \begin{pmatrix} u(r, c) \\ v(r, c) \end{pmatrix}$$

$$= \sum_{l=1}^{m} \sum_{h=1}^{n} \left(\sum_{p=1}^{n} \omega(r_l, c_p) - a_l \right)^2 c_h \frac{\partial \omega(r_l, c_h)}{\partial c_h}$$

$$+ 2 \sum_{l=1}^{m} \sum_{h=1}^{n} \left(\sum_{p=1}^{n} \omega(r_l, c_p) - a_l \right) \left(\sum_{p=1}^{m} \omega(r_p, c_h) - b_h \right) c_h \frac{\partial \omega(r_l, c_h)}{\partial c_h}$$

$$+ \sum_{l=1}^{m} \sum_{h=1}^{n} \left(\sum_{p=1}^{m} \omega(r_p, c_h) - b_h \right)^2 c_h \frac{\partial \omega(r_l, c_h)}{\partial c_h}$$

$$= \sum_{l=1}^{m} \sum_{h=1}^{n} \left(\left(\sum_{p=1}^{n} \omega(r_l, c_p) - a_l \right)^2 \right.$$

$$+ 2 \left(\sum_{p=1}^{n} \omega(r_1, c_p) - a_l \right) \left(\sum_{p=1}^{m} \omega(r_p, c_h) - b_h \right)$$

$$+ \left. \left(\sum_{p=1}^{m} \omega(r_p, c_h) - b_h \right)^2 \right) c_h \frac{\partial \omega(r_l, c_h)}{\partial c_h}$$

$$= \sum_{l=1}^{m} \sum_{h=1}^{n} \left(\sum_{p=1}^{n} \omega(r_l, c_p) - a_l \right.$$

$$+ \left. \sum_{p=1}^{m} \omega(r_p, c_h) - b_h \right)^2 c_h \frac{\partial \omega(r_l, c_h)}{\partial c_h}.$$

$$(6.39)$$

Let

$$\phi_l = \sum_{p=1}^{n} \omega(r_1, c_p) - a_l$$

and

$$\varphi_h = \sum_{p=1}^{m} \omega(r_p, c_h) - b_h.$$

We now show that if

$$\phi_l + \phi_h = 0, \quad l = 1, 2, \ldots, m, \quad h = 1, 2, \ldots, n, \quad (6.40)$$

then

$$\phi_l = 0, \quad l = 1, 2, \ldots, m, \tag{6.41}$$

and

$$\phi_h = 0, \quad h = 1, 2, \ldots, n. \tag{6.42}$$

From $\phi_l + \varphi_h = 0$, $h = 1, 2, \ldots, n$, we have that φ_h, $h = 1, 2, \ldots, n$, are equal. From $\phi_l + \varphi_l = 0$, $h = 1, 2, \ldots, m$, we have that ϕ_l, $l = 1, 2, \ldots, m$, are equal. Let $\phi_l = \phi$ for $l = 1, 2, \ldots, m$, and let $\varphi_h = \varphi$ for $h = 1, 2, \ldots, n$. Then

$$\phi + \varphi = 0. \tag{6.43}$$

Using $\sum_{i=1}^{m} a_i = \sum_{j=1}^{n} b_j$, we can observe that

$$
\begin{aligned}
\sum_{l=1}^{m} \phi_l &= \sum_{l=1}^{m} \left(\sum_{p=1}^{n} \omega(r_l, c_p) - a_l \right) \\
&= \sum_{p=1}^{n} \left(\sum_{l=1}^{m} \omega(r_l, c_p) - b_p \right) \\
&= \sum_{h=1}^{n} \left(\sum_{p=1}^{m} \omega(r_p, c_h) - b_h \right) \\
&= \sum_{h=1}^{n} \varphi_h.
\end{aligned}
\tag{6.44}
$$

Thus

$$m\phi = n\varphi. \tag{6.45}$$

From $\phi + \varphi = 0$ and $m\phi = n\varphi$ it is obvious that

$$\phi = \varphi = 0. \tag{6.46}$$

Therefore, when $(u(r, c), v(r, c)) \neq 0$, at least one of $\phi_l + \varphi_h$, $l = 1, 2, \ldots, m$, $h = 1, 2, \ldots, n$, is not equal to zero. Then, using (6.39), we obtain

$$\frac{dE(r, c)}{dt} = \nabla E(r, c)^T \begin{pmatrix} u(r, c) \\ v(r, c) \end{pmatrix} < 0, \tag{6.47}$$

since

$$c_h \frac{\partial \omega(r_l, c_h)}{\partial c_h} < 0.$$

Thus $(u(r, c)^T, v(r, c)^T)^T$ is a descent direction of $E(r, c)$. Hence conclusion follows in accordance with [35]. $\square$

We will further prove that procedure (6.15)–(6.16) converges to the solution of $\min_{\{r,\,c\}}\{E(r,c)\}$ or to the solution of (6.10). To do so, we prove that for $i=1,\ldots,m$, no subsequence of r_i^k, $k=0,1,\ldots$, and c_i^k, $k=0,1,\ldots$, approaches zero or infinity, respectively, and (r^k,c^k) converges to a solution $(r^*,c^*)>0$ of $\min_{r,c} E(r,c)$ (i.e., $E(r^*,c^*)=0$).

Theorem 3. *For $i=1,\ldots,m$, no subsequence of r_i^k, $k=0,1,\ldots$, and c_i^k, $k=0,1,\ldots$, approaches zero or infinity, respectively, and (r^k,c^k) converges to some $(r^*,c^*)>0$ satisfying $E(r^*,c^*)=0$.*

Proof. Suppose that $2\le m$ and $2\le n$ and that Problem (I) has a strictly feasible solution without loss of generality. Note that

$$\omega(r_i,c_i)=\begin{cases} \dfrac{1}{r_ic_j\alpha_{ij}(x)} & \text{if } 1\le j\le p_i, \\[2mm] \dfrac{u_{ij}}{1+r_ic_j\alpha_{ij}(x)} & \text{if } p_i+1\le j\le n, \end{cases}$$

$i=1,\ldots,m,\ j=1,\ldots,n$, and

$$0<\left(r^k,c^k\right),$$

$k=1,2,\ldots$. Suppose that for some i, a subsequence of r_i^k, $k=1,2,\ldots$, approaches zero. Then we must have that the entire sequence of r_i^k, $k=1,2,\ldots$, approaches zero. This implies that as $k\to\infty$, r_l^k approaches zero for $l=1,2,\ldots,m$, and c_h^k approaches infinity for $h=1,2,\ldots,n$. Consider the function

$$g(r,c)=\sum_{p=1}^{m}\ln r_p-\sum_{p=1}^{n}\ln c_p.$$

We have that $g(r^k,c^k)$ approaches $-\infty$ as $k\to\infty$. Computing the gradient of $g(r,c)$, we get

$$\nabla g(r,c)=\left(\frac{1}{r_1},\ldots,\frac{1}{r_m},-\frac{1}{c_1},\ldots,-\frac{1}{c_n}\right)^{\mathsf{T}}.$$

Thus, using $\sum_{i=1}^{m} a_i = \sum_{j=1}^{n} b_j$, we obtain

$$\nabla g(r, c)^{\top} \begin{pmatrix} u(r, c) \\ v(r, c) \end{pmatrix}$$

$$= \sum_{i=1}^{m} \left(\sum_{p=1}^{n} \omega(r_i, c_p) - a_i \right)$$

$$+ \sum_{j=1}^{n} \left(\sum_{p=1}^{m} \omega(r_p, c_j) - b_j \right)$$

$$= 0,$$

which means that $(u(r, c), v(r, c))^{\top}$ is perpendicular to the gradient of $g(r, c)$ at $(r, c) > 0$. Since $r^{k+1} = r^k + \mu_k u(r^k, c^k)$ and $c^{k+1} = c^k + \mu_k v(r^k, c^k)$, we have that $g(r^k, c^k)$ cannot approach $-\infty$ as $k \to \infty$, which yields a contradiction. So, no subsequence of r_i^k, $k = 1, 2, \ldots$, approaches zero for $i = 1, 2, \ldots, m$. Similarly, we can prove that no subsequence of c_j^k, $k = 1, 2, \ldots$, approaches zero for $j = 1, 2, \ldots, n$. From these results we can easily derive that no subsequence of r_i^k, $k = 1, 2, \ldots$, approaches infinity for $i = 1, 2, \ldots, m$, and that no subsequence of c_j^k, $k = 1, 2, \ldots$, approaches infinity for $j = 1, 2, \ldots, n$.

Note that

$$\nabla s(r^k, c^k)^{\top} \begin{pmatrix} u(r^k, c^k) \\ v(r^k, c^k) \end{pmatrix}$$

approaches zero as $k \to \infty$. Since (r^k, c^k), $k = 0, 1, \ldots$, are bounded, using (6.39), we get that $(u(r^k, c^k), v(r^k, c^k))$ approaches zero as $k \to \infty$.

Observe that

$$\|r^{k+1} - r^k\|_2 = \mu_k \|u(r^k, c^k)\|_2$$

and

$$\|c^{k+1} - c^k\|_2 = \mu_k \|v(r^k, c^k)\|_2.$$

Thus, as $k \to \infty$, (r^k, c^k) converges to some $(r^*, c^*) > 0$ satisfying $E(r^*, c^*) = 0$. The proof is completed.   □

The above shows that our method is convergent with stable performance. We will demonstrate this favorable property using numerical simulation. Naturally, a further concern would be the global convergence capability of the proposed deterministic annealing neural network algorithm. However, global convergence analysis of the annealing algorithm has long been a key yet unsolved issue. Although some relevant work (see, e.g., [36] and [37]) has shed some light on this issue, a systematic and complete analysis result still deserves further research efforts.

6.4 Numerical results

In this section, we use several test problems to demonstrate the efficiency of the proposed method.

These test problems are borrowed from Floudas and Pardalos [38] and generated randomly. We have programmed the algorithm in MATLAB®. All the problems have the following form:

$$
\begin{aligned}
\min \quad & f(x) = \sum_{i=1}^{m} \sum_{j=1}^{n} \left(c_{ij} x_{ij} + d_{ij} x_{ij}^2 \right) \\
\text{subject to} \quad & \sum_{j=1}^{n} x_{ij} = a_i, i = 1, \dots, m, \\
& \sum_{i=1}^{m} x_{ij} = b_j, j = 1, \dots, n, \\
& 0 \le x_{ij},
\end{aligned}
\tag{6.48}
$$

where $d_{ij} < 0$, $\sum_{i=1}^{m} a_i = \sum_{j=1}^{n} b_j$, $0 < a_i$, and $0 < b_j$. In the implementation of the algorithm, β initially equals 100 and is reduced by a factor of $\frac{8}{10}$ when $\|d(x^q) - x^q\|_1 < 0.001$. The details of the parameters can be found in [8]. The algorithm starts with four different initial points and terminates as soon as it converges.

As a comparison, the generalized reduced gradient (GRG) algorithm and barrier algorithm (BA) have been considered. The GRG algorithm has been embedded in the software LINGO, and the BA algorithm has been embedded in the software CPLEX for nonconvex quadratic programming problems.

Example 6. This example is from Floudas and Pardalos [38], and the data are presented as follows: $m = 6$, $n = 4$, $a = (8, 24, 20, 24, 16, 12)^{\top}$, $b = (29, 41, 13, 21)^{\top}$,

$$
c = (c_{ij}) = \begin{pmatrix}
300 & 270 & 460 & 800 \\
740 & 600 & 540 & 380 \\
300 & 490 & 380 & 760 \\
430 & 250 & 390 & 600 \\
210 & 830 & 470 & 680 \\
360 & 290 & 400 & 310
\end{pmatrix},
$$

and

$$d = (d_{ij}) = \begin{pmatrix} -7 & -4 & -6 & -8 \\ -12 & -9 & -14 & -7 \\ -13 & -12 & -8 & -4 \\ -7 & -9 & -16 & -8 \\ -4 & -10 & -21 & -13 \\ -17 & -9 & -8 & -4 \end{pmatrix}.$$

The algorithm converges to the optimal solution x^* after 61 iterations on average, at which the objective function equals 15639. The same problem has been solved by GRG and BA algorithms. The objective function equals 20040 by GRG and 15990 by BA. At this time the performance of NNAA is 21.96% better than that of GRG and 2.20% better than that of BA.

$$x^* = (x_{ij}^*) = \begin{pmatrix} 6 & 2 & 0 & 0 \\ 0 & 3 & 0 & 21 \\ 20 & 0 & 0 & 0 \\ 0 & 24 & 0 & 0 \\ 3 & 0 & 13 & 0 \\ 0 & 12 & 0 & 0 \end{pmatrix}.$$

Figs. 6.3 and 6.4 depict the state trajectories of the network in the first two state variables and the objective function, respectively, where each trajectory starts from four different initial points. We can see from Figs. 6.3 and 6.4 that the network converges to the optimal solution x^* at any initial point.

Example 7. This example is generated randomly. The data are presented as follows:

$$m = 10, n = 5, a = (11, 16, 1, 11, 10, 19, 15, 11, 18, 12)^{\mathsf{T}},$$
$$b = (17, 3, 4, 14, 86)^{\mathsf{T}},$$

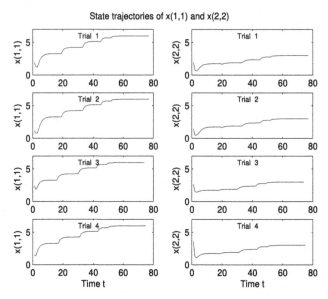

Figure 6.3 Convergent trajectories of the first two state variables from four random initial points for Example 6.

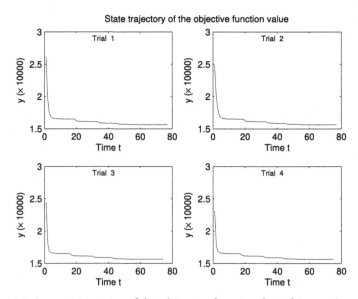

Figure 6.4 State trajectories of the objective function from four random initial points for Example 6.

$$c = (c_{ij}) = \begin{pmatrix} 23 & 54 & 54 & 34 & 8 \\ 6 & 68 & 10 & 64 & 64 \\ 69 & 2 & 66 & 77 & 89 \\ 69 & 39 & 43 & 100 & 28 \\ 94 & 8 & 71 & 38 & 45 \\ 39 & 43 & 92 & 26 & 78 \\ 53 & 70 & 77 & 99 & 49 \\ 84 & 60 & 27 & 73 & 25 \\ 4 & 94 & 6 & 76 & 28 \\ 6 & 86 & 75 & 66 & 37 \end{pmatrix},$$

and

$$d = (d_{ij}) = \begin{pmatrix} -3 & -6 & -6 & -7 & -9 \\ -6 & -4 & -10 & -8 & -5 \\ -10 & -2 & -2 & -8 & -9 \\ -10 & -10 & -9 & -11 & -4 \\ -2 & -2 & -9 & -10 & -5 \\ -10 & -6 & -9 & -3 & -6 \\ -6 & -5 & -2 & -4 & -6 \\ -6 & -4 & -1 & -5 & -4 \\ -4 & -10 & -8 & -6 & -3 \\ -11 & -6 & -10 & -7 & -3 \end{pmatrix}.$$

For four random initial points, the algorithm converges to the solution

$$x^* = (x_{ij}^*) = \begin{pmatrix} 0 & 0 & 0 & 0 & 11 \\ 5 & 2 & 4 & 4 & 1 \\ 0 & 1 & 0 & 0 & 0 \\ 0 & 0 & 0 & 0 & 11 \\ 0 & 0 & 0 & 10 & 0 \\ 0 & 0 & 0 & 0 & 19 \\ 0 & 0 & 0 & 0 & 15 \\ 0 & 0 & 0 & 0 & 11 \\ 0 & 0 & 0 & 0 & 18 \\ 12 & 0 & 0 & 0 & 0 \end{pmatrix}$$

on average 180 iterations, at which the objective function equals -5218. The same problem has been solved by GRG and BA algorithms. The objective function equals -3894 by GRG and -4764 by BA. The per-

State trajectories of x(1,1) and x(2,2)

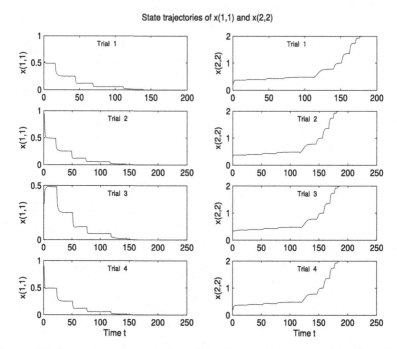

Figure 6.5 Convergent trajectories of the first two state variables from four random initial points for Example 7.

formance of NNAA is 34.00% better than that of GRG and 9.53% better than that of BA.

Figs. 6.5 and 6.6 depict the state trajectories of the network in the first two state variables and the objective function, respectively, where each trajectory starts from four different initial points. We can see from Figs. 6.5 and 6.6 that the network converges to a solution x^* at any initial point.

Example 8. This example is generated randomly. The data are presented as follows:

$$m = 20, n = 10,$$
$$a = (7, 17, 8, 15, 7, 1, 3, 11, 17, 5, 2, 2, 4, 15, 7, 2, 11, 10, 11, 10)^\top,$$
$$b = (4, 7, 18, 15, 13, 15, 8, 2, 21, 62)^\top,$$

State trajectory of the objective function value

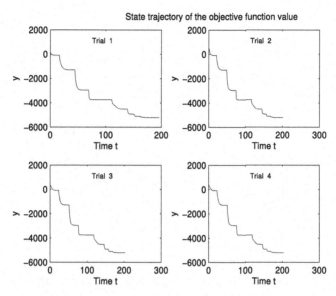

Figure 6.6 State trajectories of the objective function from four random initial points for Example 7.

$$
c = (c_{ij}) = \begin{pmatrix}
23 & 54 & 8 & 50 & 64 & 58 & 72 & 42 & 91 & 23 \\
6 & 10 & 64 & 28 & 75 & 81 & 95 & 15 & 44 & 69 \\
69 & 66 & 89 & 10 & 74 & 4 & 25 & 57 & 15 & 92 \\
69 & 43 & 28 & 96 & 101 & 54 & 19 & 26 & 96 & 26 \\
94 & 71 & 45 & 8 & 90 & 51 & 33 & 50 & 42 & 87 \\
39 & 92 & 78 & 51 & 24 & 97 & 90 & 47 & 14 & 48 \\
53 & 77 & 49 & 39 & 32 & 76 & 66 & 97 & 90 & 52 \\
84 & 27 & 25 & 29 & 36 & 56 & 16 & 14 & 10 & 61 \\
4 & 6 & 28 & 92 & 52 & 90 & 69 & 21 & 17 & 83 \\
6 & 75 & 37 & 54 & 60 & 63 & 40 & 33 & 8 & 77 \\
54 & 34 & 18 & 47 & 86 & 85 & 40 & 64 & 38 & 47 \\
68 & 64 & 50 & 95 & 42 & 17 & 51 & 14 & 26 & 96 \\
2 & 77 & 91 & 6 & 85 & 22 & 16 & 66 & 15 & 64 \\
39 & 100 & 92 & 77 & 28 & 72 & 60 & 63 & 79 & 45 \\
8 & 38 & 7 & 78 & 43 & 14 & 86 & 81 & 47 & 83 \\
43 & 26 & 91 & 84 & 55 & 10 & 60 & 26 & 36 & 70 \\
70 & 99 & 51 & 14 & 48 & 28 & 97 & 49 & 46 & 71 \\
60 & 73 & 53 & 3 & 30 & 1 & 57 & 40 & 82 & 100 \\
94 & 76 & 33 & 70 & 19 & 42 & 16 & 21 & 94 & 96 \\
86 & 66 & 100 & 88 & 16 & 4 & 99 & 4 & 66 & 86
\end{pmatrix},
$$

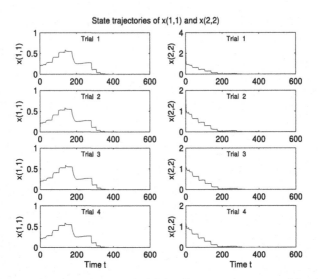

Figure 6.7 Convergent trajectories of the first two state variables from four random initial points for Example 8.

and

$$d = (d_{ij}) = \begin{pmatrix} -4 & -6 & -3 & -7 & -9 & -7 & -6 & -10 & -10 & -1 \\ -6 & -8 & -11 & -3 & -2 & -3 & -4 & -6 & -4 & -6 \\ -6 & -8 & -3 & -3 & -5 & -4 & -7 & -10 & -11 & -2 \\ -2 & -3 & -7 & -2 & -7 & -8 & -6 & -4 & -3 & -3 \\ -5 & -10 & -7 & -5 & -3 & -2 & -5 & -6 & -9 & -10 \\ -7 & -10 & -1 & -11 & -9 & -9 & -4 & -6 & -8 & -9 \\ -10 & -10 & -1 & -10 & -3 & -6 & -5 & -8 & -6 & -8 \\ -5 & -6 & -9 & -5 & -11 & -7 & -5 & -5 & -9 & -5 \\ -8 & -2 & -8 & -4 & -4 & -7 & -5 & -7 & -1 & -1 \\ -10 & -6 & -4 & -8 & -3 & -4 & -6 & -11 & -3 & -7 \\ -8 & -11 & -5 & -4 & -2 & -8 & -5 & -3 & -8 & -6 \\ -9 & -3 & -8 & -9 & -3 & -2 & -4 & -8 & -8 & -2 \\ -8 & -5 & -8 & -9 & -7 & -3 & -2 & -4 & -5 & -6 \\ -8 & -4 & -3 & -5 & -8 & -6 & -5 & -6 & -7 & -5 \\ -1 & -6 & -9 & -4 & -9 & -10 & -4 & -6 & -9 & -3 \\ -10 & -10 & -8 & -3 & -8 & -9 & -8 & -9 & -11 & -5 \\ -6 & -5 & -9 & -1 & -9 & -7 & -8 & -1 & -8 & -7 \\ -6 & -6 & -2 & -3 & -8 & -8 & -6 & -4 & -4 & -8 \\ -2 & -9 & -2 & -11 & -7 & -4 & -8 & -7 & -9 & -8 \\ -8 & -5 & -9 & -3 & -2 & -2 & -8 & -6 & -3 & -6 \end{pmatrix}.$$

On average after 409 major iterations from four random initial points, the algorithm converges to the solution

$$
x^* = \left(x_{ij}^*\right) = \begin{pmatrix}
0 & 1 & 1 & 1 & 0 & 0 & 0 & 0 & 0 & 4 \\
0 & 0 & 0 & 0 & 0 & 0 & 0 & 0 & 0 & 17 \\
0 & 0 & 0 & 0 & 0 & 0 & 0 & 0 & 8 & 0 \\
0 & 0 & 0 & 0 & 0 & 0 & 8 & 0 & 0 & 7 \\
0 & 0 & 0 & 7 & 0 & 0 & 0 & 0 & 0 & 0 \\
0 & 0 & 0 & 0 & 0 & 0 & 0 & 0 & 1 & 0 \\
0 & 0 & 0 & 3 & 0 & 0 & 0 & 0 & 0 & 0 \\
0 & 0 & 0 & 0 & 0 & 0 & 0 & 0 & 11 & 0 \\
0 & 0 & 17 & 0 & 0 & 0 & 0 & 0 & 0 & 0 \\
4 & 0 & 0 & 0 & 0 & 0 & 0 & 0 & 1 & 0 \\
0 & 2 & 0 & 0 & 0 & 0 & 0 & 0 & 0 & 0 \\
0 & 0 & 0 & 0 & 0 & 0 & 0 & 2 & 0 & 0 \\
0 & 0 & 0 & 4 & 0 & 0 & 0 & 0 & 0 & 0 \\
0 & 0 & 0 & 0 & 0 & 0 & 0 & 0 & 0 & 15 \\
0 & 2 & 0 & 0 & 0 & 5 & 0 & 0 & 0 & 0 \\
0 & 2 & 0 & 0 & 0 & 0 & 0 & 0 & 0 & 0 \\
0 & 0 & 0 & 0 & 0 & 0 & 0 & 0 & 0 & 11 \\
0 & 0 & 0 & 0 & 0 & 10 & 0 & 0 & 0 & 0 \\
0 & 0 & 0 & 0 & 11 & 0 & 0 & 0 & 0 & 0 \\
0 & 0 & 0 & 0 & 2 & 0 & 0 & 0 & 0 & 8
\end{pmatrix},
$$

at which the objective function equals -6104. The same problem has been solved by GRG and BA algorithms. The objective function equals -4064 by GRG and -5864 by BA. The performance of NNAA is 50.20% better than that of GRG and 4.09% better than that of BA.

Figs. 6.7 and 6.8 depict the state trajectories of the network in the first two state variables and the objective function, respectively, where each trajectory starts from four different initial points. We can see from Figs. 6.7 and 6.8 that the network converges to a solution x^* at any initial point.

We conduct another 10 examples generated randomly with $m = 20$ and $n = 10$. Some numerical results of the NNAA, GRG, and BA algorithms are shown in Table 6.1. The abbreviations are used in the table are as follows.

1. O_{NNAA}: the object value of the NNAA algorithm,
2. O_{GRG}: the object value of the GRG algorithm,
3. C_{GRG}: $\frac{O_{GRG}-O_{NNAA}}{|O_{GRG}|} \cdot 100\%$,

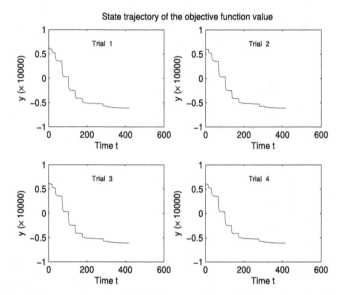

Figure 6.8 State trajectories of the objective function from four random initial points for Example 8.

4. O_{BA}: the object value of the BA algorithm,

5. C_{BA}: $\frac{O_{BA} - O_{NNAA}}{|O_{BA}|} \cdot 100\%$.

From the above 13 examples we see that the proposed algorithm performs better than the GRG and BA algorithms, which are mostly applied in real world, and the proposed algorithm is more efficient and effective.

Table 6.1 Object function values of 10 examples using NNAA, GRG, and BA.

Index	O_{NNAA}	O_{GRG}	C_{GRG}	O_{BA}	C_{BA}
1	−3709	−2310	60.56%	−3470	6.89%
2	−7096	−5204	36.36%	−6370	11.40%
3	−7626	−5102	49.47%	−6507	17.20%
4	−4923	−2890	70.35%	−4421	11.35%
5	−5442	−4262	27.69%	−5034	8.10%
6	−9840	−6701	46.84%	−8489	15.91%
7	−5633	−4139	36.10%	−4983	13.04%
8	−4430	−3279	35.10%	−4032	9.87%
9	−6926	−3937	75.92%	−5920	16.99%
10	−6487	−3968	63.48%	−5560	16.67%

6.5 Conclusions

In this chapter, to solve the minimum concave cost transportation problem efficiently, we integrate the barrier and Lagrange functions with neural network to propose a neural network annealing algorithm. The barrier parameter is controlled by annealing technique, and the solution parameters are controlled by the Lagrange function and neural network. By using this integration the proposed neural network annealing algorithm can efficiently tackle the minimum concave cost transportation problem. The experimental results also confirm the efficiency of the proposed models and algorithm for the minimum concave cost transportation problem. One important yet unsolved issue is the global convergence capability of the proposed algorithm, which will be among our future research focuses. More details about the neural network annealing algorithm can be found in [39].

References

[1] Fulya Altiparmak, Ismail Karaoglan, An adaptive tabu-simulated annealing for concave cost transportation problems, Journal of the Operational Research Society 59 (3) (2008) 331–341.

[2] Ranel E. Erickson, Clyde L. Monma, Arthur F. Veinott Jr., Send-and-split method for minimum-concave-cost network flows, Mathematics of Operations Research 12 (4) (1987) 634–664.

[3] Shu Cherng Fang, H.S.J. Tsao, Linearly-constrained entropy maximization problem with quadratic cost and its applications to transportation planning problems, Transportation Science 29 (4) (1995) 353–365.

[4] G. Gallo, C. Sodini, Concave cost minimization on networks, European Journal of Operational Research 3 (3) (1979) 239–249.

[5] Shabbir Ahmed, Qie He, Shi Li, George L. Nemhauser, On the computational complexity of minimum-concave-cost flow in a two-dimensional grid, SIAM Journal on Optimization 26 (4) (2016) 2059–2079.

[6] Shuvomoy Das Gupta, Lacra Pavel, On seeking efficient Pareto optimal points in multi-player minimum cost flow problems with application to transportation systems, arXiv preprint, arXiv:1805.11750, 2018.

[7] G.M. Guisewite, Network problems, in: Handbook of Global Optimization, Springer, 1995, pp. 609–648.

[8] Chuangyin Dang, Yabin Sun, Yuping Wang, Yang Yang, A deterministic annealing algorithm for the minimum concave cost network flow problem, Neural Networks 24 (7) (2011) 699–708.

[9] Leon S. Lasdon, Allan D. Waren, Arvind Jain, Margery Ratner, Design and testing of a generalized reduced gradient code for nonlinear programming, Technical report, Stanford Univ. CA Systems Optimization Lab, 1976.

[10] John M. Mellor-Crummey, Michael L. Scott, Algorithms for scalable synchronization on shared-memory multiprocessors, ACM Transactions on Computer Systems (TOCS) 9 (1) (1991) 21–65.

[11] John J. Hopfield, David W. Tank, Neural computation of decisions in optimization problems, Biological Cybernetics 52 (3) (1985) 141–152.

[12] John A. Hertz, Introduction to the Theory of Neural Computation, CRC Press, 2018.

[13] Qingshan Liu, Chuangyin Dang, Jinde Cao, A novel recurrent neural network with one neuron and finite-time convergence for k-winners-take-all operation, IEEE Transactions on Neural Networks 21 (7) (2010) 1140–1148.

[14] Xiaolin Hu, Bo Zhang, A new recurrent neural network for solving convex quadratic programming problems with an application to the k-winners-take-all problem, IEEE Transactions on Neural Networks 20 (4) (2009) 654–664.

[15] Xiaolin Hu, Jun Wang, An improved dual neural network for solving a class of quadratic programming problems and its k-winners-take-all application, IEEE Transactions on Neural Networks 19 (12) (2008) 2022.

[16] Xing He, Chuandong Li, Tingwen Huang, Chaojie Li, Junjian Huang, A recurrent neural network for solving bilevel linear programming problem, IEEE Transactions on Neural Networks and Learning Systems 25 (4) (2014) 824–830.

[17] Xing He, Tingwen Huang, Junzhi Yu, Chuandong Li, Chaojie Li, An inertial projection neural network for solving variational inequalities, IEEE Transactions on Cybernetics 47 (3) (2017) 809–814.

[18] Ueli Rutishauser, Jean-Jacques Slotine, Rodney J. Douglas, Solving constraint-satisfaction problems with distributed neocortical-like neuronal networks, Neural Computation 30 (5) (2018) 1359–1393.

[19] K.M. Lau, S.M. Chan, L. Xu, Comparison of the Hopfield scheme to the hybrid of Lagrange and transformation approaches for solving the traveling salesman problem, in: Intelligence in Neural and Biological Systems, 1995. INBS'95, Proceedings, First International Symposium on, IEEE, 1995, pp. 209–216.

[20] L. Xu, Combinatorial optimization neural nets based on a hybrid of Lagrange and transformation approaches, in: Proceedings of World Congress on Neutral Networks, 1994, pp. 399–404.

[21] Brandon Amos, Lei Xu, J. Zico Kolter, Input convex neural networks, arXiv preprint, arXiv:1609.07152, 2016.

[22] Van den Berg, Neural Relaxation Dynamics, PhD thesis, Erasmus University of Rotterdam, 1996.

[23] Jagat Narain Kapur, Maximum-Entropy Models in Science and Engineering, John Wiley & Sons, 1989.

[24] Gexiang Zhang, Haina Rong, Ferrante Neri, Mario J. Pérez-Jiménez, An optimization spiking neural P system for approximately solving combinatorial optimization problems, International Journal of Neural Systems 24 (05) (2014) 1440006.

[25] Yunpeng Wang, Long Cheng, Zeng-Guang Hou, Junzhi Yu, Min Tan, Optimal formation of multirobot systems based on a recurrent neural network, IEEE Transactions on Neural Networks and Learning Systems 27 (2) (2016) 322–333.

[26] Irwan Bello, Hieu Pham, Quoc V. Le, Mohammad Norouzi, Samy Bengio, Neural combinatorial optimization, arXiv preprint, arXiv:1611.09940, 2018.

[27] Xu Yang, Hong Qiao, Zhiyong Liu, An algorithm for finding the most similar given sized subgraphs in two weighted graphs, IEEE Transactions on Neural Networks and Learning Systems 29 (7) (2018) 3295–3300.

[28] Richard Durbin, David Willshaw, An analogue approach to the travelling salesman problem using an elastic net method, Nature 326 (6114) (1987) 689.

[29] A. Cochocki, Rolf Unbehauen, Neural Networks for Optimization and Signal Processing, John Wiley & Sons, Inc., 1993.

[30] Xiangsun Zhang, Neural Networks in Optimization, vol. 46, Springer Science & Business Media, 2013.

[31] Simone Scardapane, Danilo Comminiello, Amir Hussain, Aurelio Uncini, Group parse regularization for deep neural networks, Neurocomputing 241 (2017) 81–89.

[32] Ramasamy Saravanakumar, Muhammed Syed Ali, Choon Ki Ahn, Hamid Reza Karimi, Peng Shi, Stability of Markovian jump generalized neural networks with interval time-varying delays, IEEE Transactions on Neural Networks and Learning Systems 28 (8) (2017) 1840–1850.

[33] Gabriel Villarrubia, Juan F. De Paz, Pablo Chamoso, Fernando De la Prieta, Artificial neural networks used in optimization problems, Neurocomputing 272 (2018) 10–16.

[34] Yanling Wei, Ju H. Park, Hamid Reza Karimi, Yu-Chu Tian, Hoyoul Jung, Improved stability and stabilization results for stochastic synchronization of continuous-time semi-Markovian jump neural networks with time-varying delay, IEEE Transactions on Neural Networks and Learning Systems 29 (6) (2018) 2488–2501.

[35] Michael Peter Kennedy, Leon O. Chua, Neural networks for nonlinear programming, IEEE Transactions on Circuits and Systems 35 (5) (1988) 554–562.

[36] Kathryn A. Dowsland, Jonathan M. Thompson, Simulated annealing, in: Handbook of Natural Computing , 2012, pp. 1623–1655.

[37] Mahmoud H. Alrefaei, Sigrún Andradóttir, A simulated annealing algorithm with constant temperature for discrete stochastic optimization, Management Science 45 (5) (1999) 748–764.

[38] Christodoulos A. Floudas, Panos M. Pardalos, A Collection of Test Problems for Constrained Global Optimization Algorithms, vol. 455, Springer Science & Business Media, 1990.

[39] Zhengtian Wu, Hamid Reza Karimi, Chuangyin Dang, A deterministic annealing neural network algorithm for the minimum concave cost transportation problem, IEEE Transactions on Neural Networks and Learning Systems 31 (10) (2019) 4354–4366.

CHAPTER 7

An approximation algorithm for graph partitioning via deterministic annealing neural network

7.1 Introduction

The graph partitioning problem is defined as follows. Let $V = \{1, 2, \ldots, n\}$ be a set of nodes, and let E be a set of edges. In this study, we consider the undirected graph $G = (V, E)$. After defining (i, j) as an edge between nodes i and j, we assume that the graph has no isolated node. Let W be a weight matrix given by

$$W = \begin{pmatrix} 0 & w_{12} & \cdots & w_{1n} \\ w_{12} & 0 & \cdots & w_{2n} \\ \vdots & \vdots & \ddots & \vdots \\ w_{1n} & w_{2n} & \cdots & 0 \end{pmatrix},$$

where w_{ij} equals 0 when $(i, j) \notin E$, and $w_{ij} > 0$ when $(i, j) \in E$. Graph partitioning is the problem of dividing V into m disjoint sets $S_1, S_2, \ldots, S_m$, with $|S_k| = n_k$, $k = 1, 2, \ldots, m$, so that

$$\sum_{p < q} \sum_{i \in S_p, j \in S_q} w_{ij}$$

is minimized, where $|S_k|$ denotes the cardinality of S_k, $k = 1, 2, \ldots, m$. Let

$$x_{ik} = \begin{cases} 1 & \text{if } i \in S_k, \\ 0 & \text{otherwise,} \end{cases}$$

$i = 1, 2, \ldots, n$, $k = 1, 2, \ldots, m$, and

$$x = (x_{11}, x_{12}, \ldots, x_{1m}, \ldots, x_{n1}, x_{n2}, \ldots, x_{n*m}).$$

Integer Optimization and Its Computation in Emergency Management
https://doi.org/10.1016/B978-0-32-395203-3.00012-5

Then the graph partitioning problem can be described in the following form:

$$
\begin{aligned}
\min \quad & \tfrac{1}{2} \sum_{i=1}^{n} \sum_{j=1}^{n} w_{ij} \sum_{p \neq q} x_{ip} x_{jq} \\
\text{subject to} \quad & \sum_{j=1}^{m} x_{ij} = 1, \ i = 1, 2, \ldots, n, \\
& \sum_{i=1}^{n} x_{ij} = n_j, \ j = 1, 2, \ldots, m, \\
& x_{ij} \in \{0, 1\}, \ i = 1, 2, \ldots, n, \ j = 1, 2, \ldots, m.
\end{aligned}
\tag{7.1}
$$

This problem has many different applications in areas ranging from engineering to computational science. In [1,2] a provably good VLSI layout was designed successfully by graph partitioning heuristics. Image segmentation was mapped into a graph partitioning problem in [3], and it was solved in the graph partitioning formulation. In computational science the numerical solution of a boundary values problem is equal to a graph partitioning problem [4], and some boundary values problems have been solved using graph partitioning methods.

A certain number of methods have been developed to settle graph partitioning in the past few years. The Kernighan–Lin method (KLM) for partitioning a graph [5,6], which has dominated applications in industry, is a heuristic method for which some test experiments were conducted by the authors in [5]. The optimal partitioning of a graph may be nonunique, and even if it was, the way to it from any given initial partitioning may also be nonunique. Therefore the interchanging way defined by KLM has only local view. Nonetheless, KLM is widely applied in practice for easy implementation. Meanwhile, the paper [7] introduced a new graph partitioning method to deal with heterogeneous environments. The authors in [8] investigated a variant of graph partitioning. In this work the capacity constraints of the variant of the graph partitioning were imposed on the clusters, and the authors developed several compact linearized models of the problem. After considering probability models, an estimation of a distribution algorithm was developed in [9], and the graph partitioning was considered a research example to be solved by this algorithm. A memetic algorithm [10] was also developed for graph partitioning, and a local search strategy was established. To formulate a sports team realignment, [11] developed a balanced k-way partitioning with weight constraints and a method of obtaining an optimal solution. On the basis of greedy construction, a multilevel graph bipartitioning algorithm [12] was developed. The metis method (MM) [13,14] was developed on the basis of investigating the effectiveness of numerous selections for three phases, namely, coarsening, partition of the coarsest graph, and refinement. In MM, vertices on the boundaries of the partitioning are considered for interchanging or reducing computational cost. Furthermore, MM is a local optimization technique. The limitation of this method is that the starting points have a significant effect on its performance. Therefore it is commonly used as a local refinement device under the assumption that a good starting point is already available. Meanwhile, the recursive spectral partitioning (RSP) method and recursive geometric spectral partitioning (RGSP) [15] are implemented after dividing an irregular mesh into equal-sized pieces with a few interconnecting edges, and both methods have been used in practice. The idea behind these techniques is a transform of the combinatorial problem into a continuous one, given that the continuous problem already has powerful solution techniques. However, the solution obtained by these two methods cannot be validated in terms of optimality. Besides, these techniques are not very stable to some extent and sometimes result in poor partitions. RGSP is a variant of RSP, and they have similar characteristic behaviors. However, the former can obtain a better solution than the latter under the same conditions. Approximated algorithms proposed for the graph partitioning problem were reported in [16]. Other studies about graph partitioning algorithms can be obtained from [17,18]. Although considerable effort has been exerted for the graph partitioning problem, the number of algorithms that can effectively solve it remains small because of the characteristics of NP-hard combinatorial optimization problems.

Recently, new methods have been developed for this kind of optimization problems. Neural networks, which originated in [19], are one of the most efficient methods. This method and its extended algorithms have been used for a variety of optimization problems, such as bilevel linear programming [20], regression [21], class imbalance [22], and large-scale labeled graphs [23]. Dang et al. developed some algorithms based on neural networks and applied these algorithms to such problems as the traveling salesman problem [24], minimum concave cost network flow problem [25], linearly constrained nonconvex quadratic minimization problem [26], max-bisection problem [27], and min-bisection problem [28]. Some neural network algorithms have been extended to estimate the daily global solar radiation [29] and to predict evaporation [30]. A systematic study of neural network algorithms for this kind of problems was provided in [31,32]. Moreover, an elastic network combinatorial optimization algorithm was developed in [33], and other optimization neural networks can be found in the literature, for instance, [34,35]. To the best of our knowledge, however, only a few works have dealt with the use of neural networks in graph partitioning problems.

The main contribution of this study is developing a deterministic annealing neural network method that aims to obtain an approximated solution of graph partitioning. This deterministic annealing neural network (DANN) algorithm is a continuation method that attempts to identify a high-quality solution by following a path of minimum points of a barrier problem as the barrier parameter is reduced from a sufficiently large positive number to 0. The proof of the global convergence of the iterative procedure is also given in this study. Numerical results show effectiveness of the proposed method.

This chapter is organized as follows. In Section 7.2, we show that a simple continuous relaxation problem yields an equivalent continuous optimization problem to the graph partitioning and derive some important properties. In Section 7.3, we present the deterministic annealing neural network algorithm and its convergence result. In Section 7.4, we verify the global convergence of the iterative procedure and give some theorems and proofs. The numerical results are then reported in Section 7.5. The findings show that our method is effective. Finally, some concluding remarks are presented in Section 7.6.

7.2 Entropy-type Barrier function

In this section, we develop an equivalent continuous optimization problem to the graph partitioning based on a simple continuous relaxation problem. Moreover, we obtain some important properties of the equivalent problem.

For any given $\rho \geq 0$, formulation (7.1) is equivalent to the following formulation:

$$
\begin{aligned}
\min \quad & f(x) = \frac{1}{2} \sum_{i=1}^{n} \sum_{j=1}^{n} w_{ij} \sum_{p \neq q} x_{ip} x_{jq} - \frac{1}{2} \rho \sum_{i=1}^{n} \sum_{j=1}^{m} x_{ij}^2 \\
\text{subject to} \quad & \sum_{j=1}^{m} x_{ij} = 1, \ i = 1, 2, \ldots, n, \\
& \sum_{i=1}^{n} x_{ij} = n_j, \ j = 1, 2, \ldots, m, \\
& x_{ij} \in \{0, 1\}, \ i = 1, 2, \ldots, n, \ j = 1, 2, \ldots, m.
\end{aligned}
\tag{7.2}
$$

The continuous relaxation of (7.2) yields (7.3) in a process named continuous relaxation formulation:

$$
\begin{aligned}
\min \quad & f(x) = \frac{1}{2} \sum_{i=1}^{n} \sum_{j=1}^{n} w_{ij} \sum_{p \neq q} x_{ip} x_{jq} - \frac{1}{2} \rho \sum_{i=1}^{n} \sum_{j=1}^{m} x_{ij}^2 \\
\text{subject to} \quad & \sum_{j=1}^{m} x_{ij} = 1, \ i = 1, 2, \ldots, n, \\
& \sum_{i=1}^{n} x_{ij} = n_j, \ j = 1, 2, \ldots, m, \\
& 0 \leq x_{ij}, \ i = 1, 2, \ldots, n, \ j = 1, 2, \ldots, m.
\end{aligned}
\tag{7.3}
$$

When ρ is sufficiently large, (7.3) has an integer optimal solution. Thus (7.3) is equivalent to (7.2) when ρ is large enough. The size of ρ has an important impact on the quality of the solution obtained using DANN algorithm. The size of ρ should be as small as possible.

Let

$$P = \left\{ x \in R^{n \times m} \; \middle| \; \begin{array}{l} \sum_{j=1}^{m} x_{ij} = 1, \; i = 1, 2, \ldots, n, \\ \sum_{i=1}^{n} x_{ij} = n_j, \; j = 1, 2, \ldots, m, \\ 0 \leq x_{ij}, \; i = 1, 2, \ldots, n, \; j = 1, 2, \ldots, m. \end{array} \right\}.$$

Afterward, P is bounded and is the feasible region of (7.3). For the solution of (7.3), a barrier term $x_{ij} \ln x_{ij} - x_{ij}$ can be constructed to incorporate $0 \leq x_{ij}$ into the objective function of (7.3). Then we can obtain the barrier formulation

$$\begin{aligned} \min \quad & e(x; \beta) = f(x) + \beta \sum_{i=1}^{n} \sum_{j=1}^{m} (x_{ij} \ln x_{ij} - x_{ij}) \\ \text{subject to} \quad & \sum_{j=1}^{m} x_{ij} = 1, \; i = 1, 2, \ldots, n, \\ & \sum_{i=1}^{n} x_{ij} = n_j, \; j = 1, 2, \ldots, m, \end{aligned} \tag{7.4}$$

where β stands for a positive barrier parameter that behaves as the temperature in the deterministic annealing algorithm. The barrier term was first used as an energy function by Hopfield [36] and has been widely used in the literature.

A solution of (7.3) is obtained from the solution of (7.4) as $\beta \to 0$.

From $f(x)$, we obtain

$$\frac{\partial f(x)}{\partial x_{ij}} = \sum_{p=1}^{n} w_{ip} \sum_{q \neq j} x_{pq} - \rho x_{ij}.$$

Clearly, $\frac{\partial f(x)}{\partial x_{ij}}$ is bounded on any bounded set, given that $\frac{\partial f(x)}{\partial x_{ij}}$ is continuous on $R^{n \times m}$. Let $b(x) = \sum_{i=1}^{n} \sum_{j=1}^{m} (x_{ij} \ln x_{ij} - x_{ij})$. Then $e(x; \beta) = f(x) + \beta b(x)$. According to $b(x)$, we obtain $\frac{\partial b(x)}{\partial x_{ij}} = \ln x_{ij}$. Thus $\lim_{x_{ij} \to 0^+} \frac{\partial b(x)}{\partial x_{ij}} = -\infty$. Using $\frac{\partial e(x;\beta)}{\partial x_{ij}} = \frac{\partial f(x)}{\partial x_{ij}} + \beta \frac{\partial b(x)}{\partial x_{ij}}$ and $\frac{\partial f(x)}{\partial x_{ij}}$ being bounded on any bounded set, we obtain $\lim_{x_{ij} \to 0^+} \frac{\partial e(x;\beta)}{\partial x_{ij}} = -\infty$. Therefore, if a local minimum solution of (7.4) is x^*, then x^* should be an interior point of P. Let

$$L(x, \lambda, \gamma) = e(x; \beta) + \sum_{i=1}^{n} \lambda_i \left(\sum_{j=1}^{m} x_{ij} - 1 \right) + \sum_{j=1}^{m} \gamma_j \left(\sum_{i=1}^{n} x_{ij} - n_j \right),$$

where, $\lambda = (\lambda_1, \lambda_2, \ldots, \lambda_n)^\top$ and $\gamma = (\gamma_1, \gamma_2, \ldots, \gamma_m)^\top$. When x^* is a minimum solution of (7.4), x^* should be an interior point of P, and there exist λ^* and γ^* satisfying the necessary condition of first-order optimality

$$\nabla_x L(x^*, \lambda^*, \gamma^*) = 0,$$
$$\sum_{j=1}^m x_{ij}^* = 1, \ i = 1, 2, \ldots, n,$$
$$\sum_{i=1}^n x_{ij}^* = n_j, \ j = 1, 2, \ldots, m,$$

where

$$\nabla_x L(x, \lambda, \gamma) = \left(\frac{\partial L(x, \lambda, \gamma)}{\partial x_{11}}, \frac{\partial L(x, \lambda, \gamma)}{\partial x_{12}}, \ldots, \frac{\partial L(x, \lambda, \gamma)}{\partial x_{1m}}, \right.$$
$$\left. \ldots, \frac{\partial L(x, \lambda, \gamma)}{\partial x_{n1}}, \frac{\partial L(x, \lambda, \gamma)}{\partial x_{n2}}, \ldots, \frac{\partial L(x, \lambda, \gamma)}{\partial x_{n*m}} \right)^\top$$

with

$$\frac{\partial L(x, \lambda, \gamma)}{\partial x_{ij}} = \frac{\partial f(x)}{\partial x_{ij}} + \lambda_i + \gamma_j + \beta \ln x_{ij}.$$

Let β_k, $k = 0, 1, \ldots$, be a series of positive values such that $\beta_0 > \beta_1 > \cdots$ and $\lim_{k \to \infty} \beta_k = 0$. Let $x(\beta_k)$ represent a global minimum solution of (7.4), where $\beta = \beta_k$ for $k = 0, 1, \ldots$. From the standard view given in Minoux [37] we can derive the following theorem.

Theorem 4. *For $k = 0, 1, \ldots$, $f(x(\beta_k)) \geq f(x(\beta_{k+1}))$, where $x(\beta_k)$ is an arbitrary limit value, $k = 0, 1, \ldots$, will be a global minimum solution of continuous relaxation formulation (7.3).*

According to this theorem, if we can obtain a global minimum solution of barrier formulation (7.4), then we can obtain a solution of continuous relaxation formulation (7.3), and this solution is a global minimum.

Theorem 5. *For $k = 0, 1, \ldots$, x^k stands for a local minimum solution of barrier formulation (7.4), where $\beta = \beta_k$. Assume that for each limit value x^* of x^k, $k = 0, 1, \ldots$, there are no $\lambda^* \in R^n$ and $\gamma^* \in R^m$ satisfying $\frac{\partial f(x^*)}{\partial x_{ij}} + \lambda_i^* + \gamma_j^* = 0$, $i = 1, 2, \ldots, n, j = 1, 2, \ldots, m$. Then each limit value of x^k, $k = 0, 1, \ldots$, is at least a local minimum solution of continuous relaxation formulation (7.3).*

Proof. Given that x^k, $k = 0, 1, \ldots$, falls into the bounded set P, a convergent subsequence can be extracted. Let x^{k_q}, $q = 0, 1, \ldots$, be a convergent subsequence of x^k, $k = 0, 1, \ldots$. Denote $x^* = \lim_{q \to \infty} x^{k_q}$.

Since x^{k_q} is a local minimum solution of (7.4) and $\beta = \beta_{k_q}$, we obtain that there are $\lambda^{k_q} = (\lambda_1^{k_q}, \lambda_2^{k_q}, \dots, \lambda_n^{k_q})^\top$ and $\gamma^{k_q} = (\gamma_1^{k_q}, \gamma_2^{k_q}, \dots, \gamma_m^{k_q})^\top$ satisfying

$$\frac{\partial f(x^{k_q})}{\partial x_{ij}} + \lambda_i^{k_q} + \gamma_j^{k_q} + \beta_{k_q} \ln x_{ij}^{k_q} = 0,$$

$i = 1, 2, \dots, n, j = 1, 2, \dots, m$. Therefore

$$\frac{\partial f(x^*)}{\partial x_{ij}} = \lim_{q \to \infty} \frac{\partial f(x^{k_q})}{\partial x_{ij}}$$

$$= -\lim_{q \to \infty} (\lambda_i^{k_q} + \gamma_j^{k_q} + \beta_{k_q} \ln x_{ij}^{k_q}), \tag{7.5}$$

$i = 1, 2, \dots, n, j = 1, 2, \dots, m$. Let x be an interior point of P. Then

$$\sum_{i=1}^{n} \sum_{j=1}^{m} (x_{ij} - x_{ij}^{k_q}) \frac{\partial f(x^{k_q})}{\partial x_{ij}}$$
$$= -(\sum_{i=1}^{n} \lambda_i^{k_q} \sum_{j=1}^{m} (x_{ij} - x_{ij}^{k_q}) + \sum_{j=1}^{m} \gamma_j^{k_q} \sum_{i=1}^{n} (x_{ij} - x_{ij}^{k_q})$$
$$+ \beta_{k_q} \sum_{i=1}^{n} \sum_{j=1}^{m} (x_{ij} - x_{ij}^{k_q}) \ln x_{ij}^{k_q})$$
$$= -\beta_{k_q} \sum_{i=1}^{n} \sum_{j=1}^{m} (x_{ij} - x_{ij}^{k_q}) \ln x_{ij}^{k_q}.$$

Consider $K = \{(i, j) \mid x_{ij}^* = 0\}$. Then, given any $(i, j) \notin K$,

$$\lim_{q \to \infty} \beta_{k_q} (x_{ij} - x_{ij}^{k_q}) \ln x_{ij}^{k_q} = 0.$$

Let $(i, j) \in K$. We can obtain $x_{ij} - x_{ij}^* > 0$ and $\lim_{q \to \infty} x_{ij}^{k_q} = 0$. If q is sufficiently large, then

$$\beta_{k_q} (x_{ij} - x_{ij}^{k_q}) \ln x_{ij}^{k_q} < 0.$$

According to (7.5) and the assumption, $K \neq \emptyset$, and at least one of the limits

$$\lim_{q \to \infty} \beta_{k_q} \ln x_{ij}^{k_q}, \ (i, j) \in K,$$

is not equal to 0. Therefore at least one of

$$(x_{ij} - x_{ij}^*) \lim_{q \to \infty} \beta_{k_q} \ln x_{ij}^{k_q}, \ (i, j) \in K,$$

is negative. Moreover, all they are not positive. Thus

$$\sum_{i=1}^{n} \sum_{j=1}^{m} (x_{ij} - x_{ij}^{*}) \frac{\partial f(x^{*})}{\partial x_{ij}}$$
$$= \lim_{q \to \infty} \sum_{i=1}^{n} \sum_{j=1}^{m} (x_{ij} - x_{ij}^{k_q}) \frac{\partial f(x^{k_q})}{\partial x_{ij}}$$
$$= -\lim_{q \to \infty} \beta_{k_q} \sum_{i=1}^{n} \sum_{j=1}^{m} (x_{ij} - x_{ij}^{k_q}) \ln x_{ij}^{k_q}$$
$$= -\lim_{q \to \infty} \beta_{k_q} \sum_{(i,j) \in K} (x_{ij} - x_{ij}^{k_q}) \ln x_{ij}^{k_q}$$
$$= -\sum_{(i,j) \in K} (x_{ij} - x_{ij}^{*}) \lim_{q \to \infty} \beta_{k_q} \ln x_{ij}^{k_q}$$
$$> 0.$$

(7.6)

The function $f(x)$ is a quadratic, and it can be easily reformulated to matrix form $f(x) = \frac{1}{2} x^{\top} Q x$. From this matrix form we obtain $\nabla f(x) = Q x$ and

$$f(x) - f(x^{*}) = \frac{1}{2} x^{\top} Q x - \frac{1}{2} (x^{*})^{\top} Q x^{*}$$
$$= (x - x^{*})^{\top} Q x^{*} + \frac{1}{2} (x - x^{*})^{\top} Q (x - x^{*}).$$

If x becomes an interior point of P and becomes sufficiently close to x^{*}, then using (7.6), we obtain

$$f(x) - f(x^{*}) > 0$$

because $(x - x^{*})^{\top} Q x^{*} = \sum_{i=1}^{n} \sum_{j=1}^{m} (x_{ij} - x_{ij}^{*}) \frac{\partial f(x^{*})}{\partial x_{ij}} > 0$ and $\frac{1}{2} (x - x^{*})^{\top} Q (x - x^{*})$ goes to 0 twice as fast as $(x - x^{*})^{\top} Q x^{*}$ if x approaches x^{*}. Therefore x^{*} is a local minimum solution of (7.3). This completes the proof. $\square$

According to this theorem, a local minimum solution of continuous relaxation formulation (7.3) can be identified at least when a local minimum point of barrier formulation (7.4) can be obtained for a series of reduced barrier parameter values from a sufficiently large positive number to 0.

7.3 DANN algorithm for graph partitioning

In this section, we adopt the DANN method for graph partitioning. Let β be any given positive number; the necessary first-order optimality condition on (7.4) is described as follows:

$$\nabla_{x} L(x, \lambda, \gamma) = 0,$$
$$\sum_{j=1}^{m} x_{ij} = 1, \ i = 1, 2, \ldots, n,$$
$$\sum_{i=1}^{n} x_{ij} = n_j, \ j = 1, 2, \ldots, n.$$

From

$$\frac{\partial L(x, \lambda, \gamma)}{\partial x_{ij}} = \frac{\partial f(x)}{\partial x_{ij}} + \lambda_i + \gamma_j + \beta \ln x_{ij} = 0$$

we derive

$$x_{ij} = \frac{1}{e^{(\frac{\partial f(x)}{\partial x_{ij}} + \lambda_i + \gamma_j)/\beta}} = \frac{1}{e^{\lambda_i/\beta} e^{\gamma_j/\beta} e^{\frac{\partial f(x)}{\partial x_{ij}}/\beta}}.$$

Let $s_i = e^{\lambda_i/\beta}$ and $t_j = e^{\gamma_j/\beta}$. Then

$$x_{ij} = \frac{1}{s_i t_j e^{\frac{\partial f(x)}{\partial x_{ij}}/\beta}}. \tag{7.7}$$

Let $t = (t_1, t_2, \ldots, t_m)^\top$ and $s = (s_1, s_2, \ldots, s_n)^\top$. Then

$$\lambda = \beta \ln s = \beta (\ln s_1, \ln s_2, \ldots, \ln s_n)^\top \text{ and } \gamma = \beta \ln t = \beta (\ln t_1, \ln t_2, \ldots, \ln t_m)^\top.$$

Let

$$d_{ij}(x, s, t) = \frac{1}{s_i t_j e^{\frac{\partial f(x)}{\partial x_{ij}}/\beta}},$$

$i = 1, 2, \ldots, n, j = 1, 2, \ldots, m,$ and

$$d(x, s, t) = (d_{11}(x, s, t), d_{12}(x, s, t), \ldots, d_{1m}(x, s, t),$$
$$\ldots, d_{n1}(x, s, t), d_{n2}(x, s, t), \ldots, d_{nm}(x, s, t))^\top.$$

When $x > 0$, we can easily obtain the following:

- $\frac{\partial L(x, \lambda, \gamma)}{\partial x_{ij}} > 0$ if $d_{ij}(x, s, t) - x_{ij} < 0$;
- $\frac{\partial L(x, \lambda, \gamma)}{\partial x_{ij}} < 0$ if $d_{ij}(x, s, t) - x_{ij} > 0$;
- $\frac{\partial L(x, \lambda, \gamma)}{\partial x_{ij}} = 0$ if $d_{ij}(x, s, t) - x_{ij} = 0$;
- $(d(x, s, t) - x)^\top \nabla_x L(x, \lambda, \gamma) < 0$ if $d(x, s, t) - x \neq 0$;
- $(d(x, s, t) - x)^\top \nabla_x e(x; \beta) < 0$ if $d(x, s, t) - x \neq 0$, $\sum_{k=1}^m (d_{ik}(x, s, t) - x_{ik}) = 0, i = 1, 2, \ldots, n$, and $\sum_{k=1}^n (d_{kj}(x, s, t) - x_{kj}) = 0, j = 1, 2, \ldots, m.$

Thus $(d(x, s, t) - x)$ is a descent direction of $L(x, \lambda, \gamma)$. By substituting (7.7) into $\sum_{j=1}^m x_{ij} = 1, i = 1, 2, \ldots, n$, and $\sum_{i=1}^n x_{ij} = n_j, j = 1, 2, \ldots, m$, we obtain

$$\sum_{j=1}^m \frac{1}{s_i t_j e^{\frac{\partial f(x)}{\partial x_{ij}}/\beta}} = 1, \ i = 1, 2, \ldots, n,$$

$$\sum_{i=1}^n \frac{1}{s_i t_j e^{\frac{\partial f(x)}{\partial x_{ij}}/\beta}} = n_j, \ j = 1, 2, \ldots, m. \tag{7.8}$$

Let x be any interior point of P, and let $(s(x), t(x))$ be a positive point of (7.8). Then $d(x, s(x), t(x)) - x$ is a feasible descent direction of (7.4). Thus, given any interior point x of P, for $d(x, s, t) - x$ to be a feasible descent direction of (7.4), we must determine a positive point $(s, t) = (s(x), t(x))$ of (7.8).

Let

$$w(s, t) = \frac{1}{2}\left(\sum_{i=1}^{n}\left(\sum_{j=1}^{m}\frac{1}{s_i t_j e^{\frac{\partial f(x)}{\partial x_{ij}}/\beta}} - 1\right)^2 + \sum_{j=1}^{m}\left(\sum_{i=1}^{n}\frac{1}{s_i t_j e^{\frac{\partial f(x)}{\partial x_{ij}}/\beta}} - n_j\right)^2\right).$$

Clearly, $w(s, t) = 0$ if and only if (s, t) is a solution of (7.8). Let

$$u_i(s, t) = s_i\left(\sum_{j=1}^{m}\frac{1}{s_i t_j e^{\frac{\partial f(x)}{\partial x_{ij}}/\beta}} - 1\right), \quad i = 1, 2, \ldots, n,$$

and $u(s, t) = (u_1(s, t), u_2(s, t), \ldots, u_n(s, t))^\top$. Let

$$v_j(s, t) = t_j\left(\sum_{i=1}^{n}\frac{1}{s_i t_j e^{\frac{\partial f(x)}{\partial x_{ij}}/\beta}} - n_j\right), \quad j = 1, 2, \ldots, m,$$

and $v(s, t) = (v_1(s, t), v_2(s, t), \ldots, v_m(s, t))^\top$. We will show in the next section that a descent direction of $w(s, t)$ is $(u(s, t), v(s, t))$.

Let x be any given number. According to $(u(s, t), v(s, t))$, an iterative procedure for computing a positive solution $(s(x), t(x))$ of (7.8) is developed as follows.

Choose an arbitrary positive vector denoted by (s^0, t^0). For $k = 0, 1, \ldots$, consider

$$\begin{aligned} s^{k+1} &= s^k + \mu_k u(s^k, t^k), \\ t^{k+1} &= t^k + \mu_k v(s^k, t^k), \end{aligned} \tag{7.9}$$

where μ_k is a number in $[0, \frac{1}{\max_{1 \leq j \leq m} n_j}]$ satisfying

$$w(s^{k+1}, t^{k+1}) = \min_{\mu \in [0, \frac{1}{\max_{1 \leq j \leq m} n_j}]} w(s^k + \mu u(s^k, t^k), t^k + \mu v(s^k, t^k)).$$

Clearly, $(s^k, t^k) > 0$, $k = 0, 1, \ldots$. In the implementation of (7.9), an exact solution of $\min_{\mu \in [0, \frac{1}{\max_{1 \leq j \leq m} n_j}]} w(s^k + \mu u(s^k, t^k), t^k + \mu v(s^k, t^k))$ need not be solved when an approximate solution has the same performance. Many other ways can be used to identify μ_k [37]. We can consider a simple example; let $\mu_k \in (0, \frac{1}{\max_{1 \leq j \leq m} n_j}]$ satisfy $\mu_k \to 0$ and $\sum_{l=0}^{k} \mu_l \to \infty$ as $k \to \infty$. Our numerical tests show that (7.9) always converges if μ_k is a fixed number in $(0, \frac{1}{\max_{1 \le j \le m} n_j})$. The global convergence of (7.9) is proved in the next section.

In the above analysis, the feasible descent direction $d(x, s(x), t(x)) - x$ is obtained, and the iterative procedure (7.9) is constructed. In the next part, we develop a deterministic annealing method for approximating a solution of (7.3). The algorithm is initialized as follows.

Initialization: *Let ϵ be any given tolerance, and let β_0 be a sufficiently large positive number that makes $e(x; \beta_0)$ convex. Let $\bar{x}$ be an arbitrary point that satisfies $0 < \bar{x}_{ij} < 1$, and let (s^0, t^0) be an arbitrary positive vector. Let $\eta \in (0, 1)$ be an arbitrary positive number, where we need η to be close to 1. Given that $x = \bar{x}$, based on (7.9), we can obtain a positive solution $(s(\bar{x}), t(\bar{x}))$ of (7.8). Let $(s^0, t^0) = (s(\bar{x}), t(\bar{x}))$ and*

$$x^0 = \left(x_{11}^0, x_{12}^0, \ldots, x_{1m}^0, \ldots, x_{n1}^0, x_{n2}^0, \ldots, x_{n*m}^0\right)^{\top}$$

with $x_{ij}^0 = \dfrac{1}{s_i(\bar{x})t_j(\bar{x})e^{\frac{\partial f(\bar{x})}{\partial x_{ij}}/\beta_0}}$, $i = 1, 2, \ldots, n$, $j = 1, 2, \ldots, m$. Set $k = 0$ and $q = 0$. The algorithm starts.

The flowchart of the algorithm is given Fig. 7.1.

Step 1: Let $x = x^k$, and apply (7.9) to compute a positive solution $(s(x^k), t(x^k))$ of (7.8). Let $(s^0, t^0) = (s(x^k), t(x^k))$, and go to Step 2.

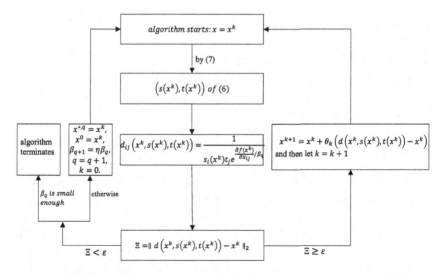

Figure 7.1 Flowchart of the DANN algorithm.

Step 2: Consider

$d(x^k, s(x^k), t(x^k)) =$
$(d_{11}(x^k, s(x^k), t(x^k)), d_{12}(x^k, s(x^k), t(x^k)), \cdots, d_{1m}(x^k, s(x^k), t(x^k)), \ldots,$
$d_{n1}(x^k, s(x^k), t(x^k)), d_{n2}(x^k, s(x^k), t(x^k)), \ldots, d_{nm}(x^k, s(x^k), t(x^k)))^\top$

with $d_{ij}(x^k, s(x^k), t(x^k)) = \dfrac{1}{s_i(x^k)t_j(x^k)e^{\frac{\partial f(x^k)}{\partial x_{ij}}/\beta_q}}$, $i = 1, 2, \ldots, n, j = 1, 2, \ldots, m.$

If $\|d(x^k, s(x^k), t(x^k)) - x^k\|_2 < \epsilon$, then do the following actions.

- The algorithm terminates if β_q is small enough.
- Otherwise, let $x^{*,q} = x^k$, $x^0 = x^k$, $\beta_{q+1} = \eta\beta_q$, $q = q + 1$, $k = 0$; then go to Step 1.

If $\|d(x^k, s(x^k), t(x^k)) - x^k\|_2 \geq \epsilon$, then do the following actions. Solve

$$x^{k+1} = x^k + \theta_k\big(d(x^k, s(x^k), t(x^k)) - x^k\big), \tag{7.10}$$

where θ_k is in $[0, 1]$ and satisfies

$$e\big(x^{k+1}; \beta_q\big) = \min_{\theta\in[0,1]} e\big(x^k + \theta\big(d(x^k, s(x^k), t(x^k)) - x^k\big); \beta_q\big).$$

Then consider $k = k + 1$ and go to Step 1.

We have $x^k > 0$, $k = 0, 1, \ldots$, and $\{x^k \mid k = 0, 1, \ldots\}$ is bound. In the implementation of the method, the exact solution of $\min_{\theta\in[0,1]} e(x^k + \theta(d(x^k, s(x^k), t(x^k)) - x^k); \beta_q)$ need not be determined, as an approximate solution has similar effects. Many other ways can be used to determine θ_k [37]. We can take a simple example. Let $\theta_k \in (0, 1]$ satisfy $\theta_k \to 0$ and $\sum_{l=0}^k \theta_l \to \infty$ as $k \to \infty$. In our numerical experiments, we apply an Armijo-type line search to determine θ_k. The method is not sensitive to the start value $\bar{x}$ because $e(x; \beta_0)$ is convex. When $x > 0$, $\nabla_x L(x, \lambda, \gamma) = 0$ if $d(x, r, t) - x = 0$. On the basis of the standard argument by Minoux [37], we easily obtain the following theorem.

Theorem 6. *Any limit solution $\beta = \beta_q$ of x^k, $k = 0, 1, \ldots$, obtained by $x^{k+1} = x^k + \theta_k(d(x^k, s(x^k), t(x^k)) - x^k)$ is a stationary solution of barrier formulation (7.4).*

It is difficult to prove that for $\beta = \beta_q$, every limit point of x^k, $k = 0, 1, \ldots$, generated by $x^{k+1} = x^k + \theta_k(d(x^k, s(x^k), t(x^k)) - x^k)$ is at least a local minimum point of barrier formulation (7.4). However, in general, it is indeed at least a local minimum point of barrier formulation (7.4). Theorem 5 implies that every limit point of $x^{*,q}$, $q = 0, 1, \ldots$, is at least a local minimum point of continuous relaxation formulation (7.3) if $x^{*,q}$ is a minimum point of barrier formulation (7.4) with $\beta = \beta_q$.

7.4 Proof of global convergence of iterative procedure

In this section, we prove that for any number x, iterative procedure (7.9) converges to a positive point (s^*, t^*) of (7.8). We first verify that a descent direction of $w(s, t)$ is $(u(s, t), v(s, t))$. Let

$$a_i(s, t) = \sum_{j=1}^{m} \frac{1}{s_i t_j e^{\frac{\partial f(x)}{\partial x_{ij}}/\beta}} - 1,$$

$i = 1, 2, \ldots, n$, and $a(s, t) = (a_1(s, t), a_2(s, t), \ldots, a_n(s, t))^\top$. Let

$$b_j(s, t) = \sum_{i=1}^{n} \frac{1}{s_i t_j e^{\frac{\partial f(x)}{\partial x_{ij}}/\beta}} - n_j,$$

$j = 1, 2, \ldots, m$, and $b(s, t) = (b_1(s, t), b_2(s, t), \ldots, b_m(s, t))^\top$. From $w(s, t)$ we obtain

$$\frac{\partial w(s,t)}{\partial s_l} = -\sum_{h=1}^{m} \frac{t_h e^{\frac{\partial f(x)}{\partial x_{lh}}/\beta}}{(s_l t_h e^{\frac{\partial f(x)}{\partial x_{lh}}/\beta})^2} \left(\sum_{p=1}^{m} \frac{1}{s_l t_p e^{\frac{\partial f(x)}{\partial x_{lp}}/\beta}} - 1 + \sum_{p=1}^{n} \frac{1}{s_p t_h e^{\frac{\partial f(x)}{\partial x_{ph}}/\beta}} - n_h \right)$$

$$= -\sum_{h=1}^{m} \frac{t_h e^{\frac{\partial f(x)}{\partial x_{lh}}/\beta}}{(s_l t_h e^{\frac{\partial f(x)}{\partial x_{lh}}/\beta})^2} (a_l(s, t) + b_h(s, t))$$

and

$$\frac{\partial w(s,t)}{\partial t_h} = -\sum_{l=1}^{n} \frac{s_l e^{\frac{\partial f(x)}{\partial x_{lh}}/\beta}}{(s_l t_h e^{\frac{\partial f(x)}{\partial x_{lh}}/\beta})^2} \left(\sum_{p=1}^{m} \frac{1}{s_l t_p e^{\frac{\partial f(x)}{\partial x_{lp}}/\beta}} - 1 + \sum_{p=1}^{n} \frac{1}{s_p t_h e^{\frac{\partial f(x)}{\partial x_{ph}}/\beta}} - n_h \right)$$

$$= -\sum_{l=1}^{n} \frac{s_l e^{\frac{\partial f(x)}{\partial x_{lh}}/\beta}}{(s_l t_h e^{\frac{\partial f(x)}{\partial x_{lh}}/\beta})^2} (a_l(s, t) + b_h(s, t)).$$

Let $\nabla w(s, t) = (\frac{\partial w(s,t)}{\partial s_1}, \frac{\partial w(s,t)}{\partial s_2}, \ldots, \frac{\partial w(s,t)}{\partial s_n}, \frac{\partial w(s,t)}{\partial t_1}, \frac{\partial w(s,t)}{\partial t_2}, \ldots, \frac{\partial w(s,t)}{\partial t_m})^\top$. The next lemma shows that $(u(s, t), v(s, t))$ should be a descent direction of $w(s, t)$.

Lemma 7. If $(s, t) > 0$ and $(u(s, t), v(s, t)) \neq 0$, then

$$\nabla w(s, t)^\top \begin{pmatrix} u(s, t) \\ v(s, t) \end{pmatrix} < 0.$$

Proof. Note that

$$u_l(s, t) = s_l \left(\sum_{p=1}^{m} \frac{1}{s_l t_p e^{\frac{\partial f(x)}{\partial x_{lp}}/\beta}} - 1 \right) = s_l a_l(s, t),$$

$l = 1, 2, \ldots, n$, and

$$v_h(s, t) = t_h \left(\sum_{p=1}^{n} \frac{1}{s_p t_h e^{\frac{\partial f(x)}{\partial x_{ph}}/\beta}} - n_h \right) = t_h b_h(s, t),$$

$h = 1, 2, \ldots, n$. Then

$$\nabla w(s, t)^{\top} \begin{pmatrix} u(s, t) \\ v(s, t) \end{pmatrix}$$

$$= -\sum_{l=1}^{n} \sum_{h=1}^{m} \frac{1}{s_l t_h e^{\frac{\partial f(x)}{\partial x_{lh}}/\beta}} \left((a_l(s, t))^2 + 2a_l(s, t)b_h(s, t) + (b_h(s, t))^2 \right) \quad (7.11)$$

$$= -\sum_{l=1}^{n} \sum_{h=1}^{m} \frac{1}{s_l t_h e^{\frac{\partial f(x)}{\partial x_{lh}}/\beta}} \left(a_l(s, t) + b_h(s, t) \right)^2.$$

We further show that if $a_l(s, t) + b_h(s, t) = 0$, $l = 1, 2, \ldots, n$, $h = 1, 2, \ldots, m$, then $a_l(s, t) = 0$, $l = 1, 2, \ldots, n$, and $b_h(s, t) = 0$, $h = 1, 2, \ldots, m$. Since $a_l(s, t) + b_h(s, t) = 0$, $h = 1, 2, \ldots, m$, it follows that $b_h(s, t)$, $h = 1, 2, \ldots, m$, are equal. Since $a_l(s, t) + b_h(s, t) = 0$, $l = 1, 2, \ldots, n$, it follows that $a_l(s, t)$, $l = 1, 2, \ldots, n$, are equal. Consider $a_l(s, t) = \phi$, $l = 1, 2, \ldots, n$, and $b_h(s, t) = \varphi$, $h = 1, 2, \ldots, m$. Then $\phi + \varphi = 0$. Since $\sum_{p=1}^{m} n_p = n$, we have

$$\sum_{l=1}^{n} a_l(s, t) = \sum_{l=1}^{n} \left(\sum_{p=1}^{m} \frac{1}{s_l t_p e^{\frac{\partial f(x)}{\partial x_{lp}}/\beta}} - 1 \right)$$

$$= \sum_{p=1}^{m} \sum_{l=1}^{n} \frac{1}{s_l t_p e^{\frac{\partial f(x)}{\partial x_{lp}}/\beta}} - n$$

$$= \sum_{p=1}^{m} \sum_{l=1}^{n} \frac{1}{s_l t_p e^{\frac{\partial f(x)}{\partial x_{lp}}/\beta}} - \sum_{p=1}^{m} n_p$$

$$= \sum_{p=1}^{m} \left(\sum_{l=1}^{n} \frac{1}{s_l t_p e^{\frac{\partial f(x)}{\partial x_{lp}}/\beta}} - n_p \right)$$

$$= \sum_{h=1}^{m} \left(\sum_{p=1}^{n} \frac{1}{s_p t_h e^{\frac{\partial f(x)}{\partial x_{ph}}/\beta}} - n_h \right)$$

$$= \sum_{h=1}^{m} b_h(s, t).$$

Thus $\phi = \varphi$. From $\phi + \varphi = 0$ and $\phi = \varphi$ we obtain $\phi = \varphi = 0$. Therefore, when $(u(s, t), v(s, t))$ is not equal to 0, at least one of $a_l(s, t) + b_h(s, t)$, $h = 1, 2, \ldots, m$, $l = 1, 2, \ldots, n$, is not 0. Therefore, according to (7.11),

$$\nabla w(s, t)^{\top} \begin{pmatrix} u(s, t) \\ v(s, t) \end{pmatrix} < 0.$$

This completes the proof.   □

From Lemma 7 we obtain the following theorem.

Theorem 7. *Let x be an arbitrary value. Any limit solution of (s^k, t^k), $k = 0, 1, \ldots$, obtained by the iterative procedure (7.9) is a positive point of the following equation:*

$$\sum_{j=1}^{m} \frac{1}{s_i t_j e^{\frac{\partial f(x)}{\partial x_{ij}}/\beta}} = 1, \ i = 1, 2, \ldots, n,$$

$$\sum_{i=1}^{n} \frac{1}{s_i t_j e^{\frac{\partial f(x)}{\partial x_{ij}}/\beta}} = n_j, \ j = 1, 2, \ldots, m.$$

Proof. Let $\bar{x}$ be an interior solution of P. Then

$$\sum_{j=1}^{m} \bar{x}_{ij} = 1, \ i = 1, 2, \ldots, n,$$

$$\sum_{i=1}^{n} \bar{x}_{ij} = n_j, \ j = 1, 2, \ldots, m,$$

$$0 < \bar{x}_{ij}, \ i = 1, 2, \ldots, n, \ j = 1, 2, \ldots, m.$$

Thus, for any $s > 0$ and $t > 0$,

$$\sum_{i=1}^{n} \ln s_i + \sum_{j=1}^{m} n_j \ln t_j = \sum_{i=1}^{n} \ln s_i \sum_{j=1}^{m} \bar{x}_{ij} + \sum_{j=1}^{m} \ln t_j \sum_{i=1}^{n} \bar{x}_{ij}$$

$$= \sum_{i=1}^{n} \sum_{j=1}^{m} \bar{x}_{ij} (\ln s_i + \ln t_j). \tag{7.12}$$

Consider

$$h(s, t) = -\sum_{i=1}^{n} \sum_{j=1}^{m} \int_{0}^{\ln s_i + \ln t_j} \frac{1}{e^v e^{\frac{\partial f(x)}{\partial x_{ij}}/\beta}} dv + \sum_{i=1}^{n} \ln s_i + \sum_{j=1}^{m} n_j \ln t_j. \tag{7.13}$$

We have

$$\nabla h(s, t) = -\left(Q(s, t)\right)^{-1} \begin{pmatrix} a(s, t) \\ b(s, t) \end{pmatrix},$$

where

$$Q(s, t) = \begin{pmatrix} s_1 & & & & & \\ & \ddots & & & & \\ & & s_n & & & \\ & & & t_1 & & \\ & & & & \ddots & \\ & & & & & t_m \end{pmatrix}.$$

Using

$$\begin{pmatrix} u(s, t) \\ v(s, t) \end{pmatrix} = Q(s, t) \begin{pmatrix} a(s, t) \\ b(s, t) \end{pmatrix},$$

we obtain that

$$\nabla h(s, t)^\top \begin{pmatrix} u(s, t) \\ v(s, t) \end{pmatrix} = -\left(a(s, t)^\top a(s, t) + b(s, t)^\top b(s, t)\right) < 0$$

when $(a(s, t), b(s, t)) \neq 0$. Thus $(u(s, t), v(s, t))$ is a descent direction of $w(s, t)$. Therefore $w(s^k, t^k)$ does not increase with k because $s^{k+1} = s^k + \mu_k u(s^k, t^k)$ and $t^{k+1} = t^k + \mu_k v(s^k, t^k)$.

By the mean value integral theorem we obtain

$$\int_0^{\ln s_i + \ln t_j} \frac{1}{e^v e^{\frac{\partial f(x)}{\partial x_{ij}}/\beta}} dv = \frac{1}{e^{v_{ij}(s,t)} e^{\frac{\partial f(x)}{\partial x_{ij}}/\beta}} (\ln s_i + \ln t_j), \qquad (7.14)$$

where $v_{ij}(s, t)$ is a positive number between 0 and $\ln s_i + \ln t_j$, which satisfies $v_{ij}(s, t) \to \infty$ as $\ln s_i + \ln t_j \to \infty$ and $v_{ij}(s, t) \to -\infty$ as $\ln s_i + \ln t_j \to -\infty$. Substituting (7.12) and (7.14) into (7.13), we obtain

$$h(s, t) = -\sum_{i=1}^{n} \sum_{j=1}^{n} \left(\frac{1}{e^{v_{ij}(s,t)} e^{\frac{\partial f(x)}{\partial x_{ij}}/\beta}} - \bar{x}_{ij} \right) (\ln s_i + \ln t_j). \qquad (7.15)$$

Let (i_0, j_0) be a pair of i and j. In the pair, the series $s_{i_0}^k t_{j_0}^k$, $k = 0, 1, \ldots$, goes to 0 or ∞. For the subsequence $s_{i_0}^{k_q} t_{j_0}^{k_q}$, $q = 0, 1, \ldots$, according to (7.15), we obtain that $h(s^{k_q}, t^{k_q}) \to \infty$ as $q \to \infty$, which contradicts the fact that $h(s^k, t^k)$ does not increase as k increases. Therefore, given any i, j, no series of $s_i^k t_j^k$, $k = 0, 1, \ldots$, goes to 0 or ∞.

Assume that there exists an index l such that a subsequence of s_l^k, $k = 0, 1, \ldots$, goes to 0. Let $s_l^{k_q}$, $q = 0, 1, \ldots$, be such a subsequence. From the above analysis we obtain that $s_i^{k_q} \to 0$ for $i = 1, 2, \ldots, n$ and $t_j^{k_q} \to \infty$ for $j = 1, 2, \ldots, n$ as $q \to \infty$. Consider

$$g(s, t) = \sum_{i=1}^{n} \ln s_i - \sum_{j=1}^{m} \ln t_j.$$

Clearly, $g(s^{k_q}, t^{k_q}) \to -\infty$ as $q \to \infty$. From $g(s, t)$ we obtain

$$\nabla g(s, t) = \left(\frac{1}{s_1}, \frac{1}{s_2}, \ldots, \frac{1}{s_n}, -\frac{1}{t_1}, -\frac{1}{t_2}, \ldots, -\frac{1}{t_m} \right)^{\top}.$$

Given that $\sum_{j=1}^{m} n_j = n$,

$$\nabla g(s, t)^{\top} \begin{pmatrix} u(s, t) \\ v(s, t) \end{pmatrix}$$

$$= \sum_{i=1}^{n} \left(\sum_{p=1}^{m} \frac{1}{s_i t_p e^{\frac{\partial f(x)}{\partial x_{ip}} / \beta}} - 1 \right) - \sum_{j=1}^{m} \left(\sum_{p=1}^{n} \frac{1}{s_p t_j e^{\frac{\partial f(x)}{\partial x_{pj}} / \beta}} - n_j \right)$$

$$= 0.$$

Then $(u(s, t), v(s, t))^{\top}$ is perpendicular to the gradient of $g(s, t)$. Therefore $g(s^{k_q}, t^{k_q})$ cannot approach $-\infty$ as $q \to \infty$ since $s^{k+1} = s^k + \mu_k u(s^k, t^k)$ and $t^{k+1} = t^k + \mu_k v(s^k, t^k)$, which causes conflict. Thus, given any i, no subsequence of s_i^k, $k = 0, 1, \ldots$, goes to 0. By the same method we can prove the following: For any j, no subsequence of t_j^k, $k = 0, 1, \ldots$, goes to 0, for any i, no subsequence of s_i^k, $k = 1, 2, \ldots$, goes to ∞, and for any j, no subsequence of t_j^k, $k = 1, 2, \ldots$, goes to ∞.

Since $w(s, t) \geq 0$ and $w(s^k, t^k)$ decreases strictly and monotonically, according to the boundedness of $\ln s_i^k$ and $\ln t_j^k$ and Lemma 7, we obtain that

$$\nabla w(s^k, t^k)^{\top} \begin{pmatrix} u(s^k, t^k) \\ v(s^k, t^k) \end{pmatrix} \to 0$$

as $k \to \infty$. Then, according to (7.11), given any i and j,

$$a_i(s^k, t^k) + b_j(s^k, t^k) \to 0$$

as $k \to \infty$. Thus $a(s^k, t^k) \to 0$ and $b(s^k, t^k) \to 0$ as $k \to \infty$. Therefore every limit point of (s^k, t^k), $k = 0, 1, \ldots$, should be a positive point of (7.8). Thus the theorem is proved. $\square$

7.5 Numerical results

In this section, we present some numerical experiments and comparisons. One hundred test problems used as benchmarks are generated randomly as follows. Let $p \in [0, 1]$ be any given number. An edge lies between two nodes if a random number following uniform distribution in $[0, 1]$ is less than or equal to p. If an edge lies between nodes i and j, then the weight w_{ij} on this edge is a random integer following the uniform distribution in $[0, 10]$; otherwise, $w_{ij} = 0$.

MATLAB® is used to program DANN algorithm. The following well-known graph partitioning methods are compared with our proposed algorithm.

1. KLM [5,6], which is widely applied in industry, is programmed in MATLAB.
2. The MM [13,14], which is based on multilevel graph partitioning, is developed as a METIS package coded by C++.
3. RSP and 4. RGSP [15] are executed by the MESHPART package coded by MATLAB.

In the implementation, $\beta = 950$ initially and is reduced by a factor of 0.95 when $\|d(x^k, s(x^k), t(x^k)) - x^k\|_2 < \epsilon$, where $\epsilon = 0.001$ denotes the accuracy of the algorithm. The smaller the ϵ, the better the solution, but the convergence time needed by the algorithm will be prolonged. For the 100 test examples, the data are as follows: $p = 0.001$, $m = 5$, $n = 100$, and the weight matrix is generated randomly under the benchmark rule. Then the numerical results of DANN algorithm and the four other methods could be obtained.

To properly compare DANN with the other methods at the same level, we consider the following definitions:

$$comp_{KLM} = \frac{KLM - DANN}{DANN} \times 100\%,$$

$$comp_{MM} = \frac{MM - DANN}{DANN} \times 100\%,$$

$$comp_{RSP} = \frac{RSP - DANN}{DANN} \times 100\%,$$

$$comp_{RGSP} = \frac{RGSP - DANN}{DANN} \times 100\%,$$

(7.16)

where $comp$ stands for the performance of DANN compared with that of each of the other methods. A positive $comp$ means that DANN is better than the compared method. The bigger the $comp$, the better the performance of DANN.

Let $comp$ be on the vertical axis, and let the 100 test examples be distributed on the horizontal axis. The comparisons of DANN with the four other methods are illustrated in Figs. 7.2–7.5. The value of the vertical axis stands for the degree of excellence. According to Equation (7.16), if the value of DANN is better than those of the other methods, the histogram will fall on the positive coordinate of the vertical axis, and vice versa.

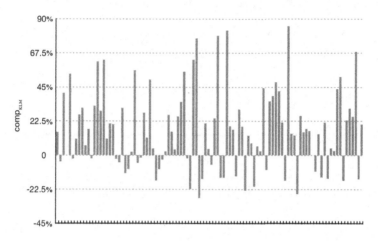

Test examples 1~100

Figure 7.2 Comparison of the object function values of DANN and KLM.

As seen in Fig. 7.2, DANN has a better object function value than KLM for 68 test examples, the same values for 2 test examples, and worse values for 30 test examples. According to the value of the vertical axis, DANN is 16.24% better than KLM on average.

As shown in Fig. 7.3, DANN has a better object function value than MM for 70 test examples, the same values for 5 test examples, and worse values for 25 test examples. According to the value of the vertical axis, DANN is 19.65% better than MM on average.

According to Fig. 7.4, DANN has a better object function value than RSP for 97 test examples, and worse values for 3 test examples. According to the value of the vertical axis, DANN is 57.15% better than RSP on average.

As displayed in Fig. 7.5, we can see that DANN method has a better object function value than RGSP for 91 test examples and worse values for 9 test examples. According to the value of the vertical axis, DANN is 34.05% better than RGSP on average.

According to Figs. 7.2–7.5 and the above analysis, although the proposed method does not perform better than the other algorithms in all

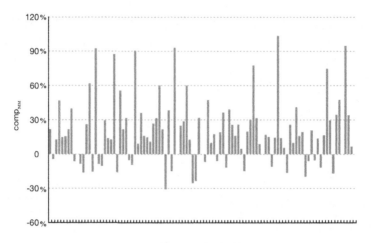

Figure 7.3 Comparison of the object function values of DANN and MM.

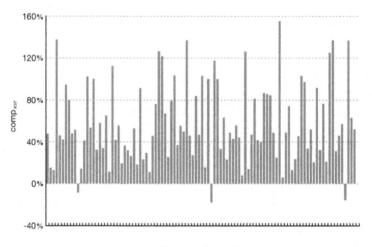

Figure 7.4 Comparison of the object function values of DANN and RSP.

cases, it is better than the others in terms of probability. However, given the common feature of annealing algorithms, the performance of DANN partly depends on the selected parameters, such as β and ϵ. The optimal values of these parameters cannot be calculated theoretically and can be obtained only from experience and experimentation. The experimentation will consume some time before computation. Besides, for algorithms

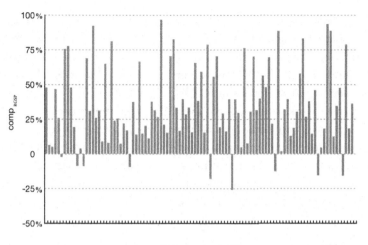

Test examples 1~100

Figure 7.5 Comparison of the object function values of DANN and RGSP.

such as annealing and genetic algorithms, we can only obtain the computational complexity in every iteration, and the total number of iterations
could not be obtained. Therefore performing a computational complexity
analysis for this algorithm is challenging.

7.6 Conclusions

In this chapter, we develop the DANN algorithm for graph partitioning,
a classical NP-hard combinatorial optimization problem. For the construction of this algorithm, we first introduce an entropy-type barrier function.
Then we prove that DANN obtains a high-quality solution by following a
path of minimum points of the barrier problem with the barrier parameter
decreasing from a sufficiently large positive number to 0. In other words,
with the barrier parameter assumed to be an arbitrary positive number, a
minimum solution of the barrier problem can be found by the algorithm in
a feasible descent direction. With a globally convergent iterative procedure,
the feasible descent direction could be obtained via renewing Lagrange
multipliers, and a distinguishing feature of it is that the upper and lower
bounds on the variables are automatically satisfied under the condition that
the step length is a value between 0 to 1. DANN is compared with four
well-known algorithms KLM, MM, RSP, and RGSP. One hundred test
examples are used to obtain comparative numerical results. On the one

hand, DANN obtains the maximum number of test examples with the best object function values. On the other hand, DANN is at least 16.30% better than the four other methods according to the average *comp*. The numerical results show that the DANN algorithm is effective. More details about the DANN algorithm can be found in [38].

In future work, some efforts about practical applications of the algorithm will be given additional attention. As practical applications, VLSI design and image segmentation will be considered. Given that DANN is a deterministic annealing algorithm, a technology report detailing the implementation of the proposed algorithm and other computing technologies will be considered to improve DANN.

References

[1] Reza H. Ahmadi, Christopher S. Tang, An operation partitioning problem for automated assembly system design, Operations Research 39 (5) (1991) 824–835.

[2] Thomas Lengauer, Combinatorial Algorithms for Integrated Circuit Layout, Springer, 2012.

[3] Jianbo Shi, Jitendra Malik, Normalized Cuts and Image Segmentation, Departmental Papers (CIS), 2000, p. 107.

[4] Erich Prisner, Graph Dynamics, vol. 338, CRC Press, 1995.

[5] Brian W. Kernighan, Shen Lin, An efficient heuristic procedure for partitioning graphs, The Bell System Technical Journal 49 (2) (1970) 291–307.

[6] Keld Helsgaun, An effective implementation of the Lin–Kernighan traveling salesman heuristic, European Journal of Operational Research 126 (1) (2000) 106–130.

[7] Ning Xu, Bin Cui, Lei Chen, Zi Huang, Yingxia Shao, Heterogeneous environment aware streaming graph partitioning, IEEE Transactions on Knowledge and Data Engineering 27 (6) (2015) 1560–1572.

[8] Viet Hung Nguyen, Michel Minoux, Improved linearized models for graph partitioning problem under capacity constraints, Optimization Methods & Software 32 (4) (2017) 892–903.

[9] Josu Ceberio, Alexander Mendiburu, Jose A. Lozano, A square lattice probability model for optimising the graph partitioning problem, in: 2017 IEEE Congress on Evolutionary Computation, IEEE, 2017, pp. 1629–1636.

[10] David Chalupa, A memetic algorithm for the minimum conductance graph partitioning problem, arXiv preprint, arXiv:1704.02854, 2017.

[11] Diego Recalde, Daniel Severín, Ramiro Torres, Polo Vaca, An exact approach for the balanced k-way partitioning problem with weight constraints and its application to sports team realignment, Journal of Combinatorial Optimization 36 (3) (2018) 916–936.

[12] Tahir Emre Kalayci, Roberto Battiti, A reactive self-tuning scheme for multilevel graph partitioning, Applied Mathematics and Computation 318 (2018) 227–244.

[13] George Karypis, Vipin Kumar, A fast and high quality multilevel scheme for partitioning irregular graphs, SIAM Journal on Scientific Computing 20 (1) (1998) 359–392.

[14] Dominique LaSalle, George Karypis, A parallel hill-climbing refinement algorithm for graph partitioning, in: 45th International Conference on Parallel Processing (ICPP), IEEE, 2016, pp. 236–241.

[15] John R. Gilbert, Gary L. Miller, Shanghua Teng, Geometric mesh partitioning: implementation and experiments, SIAM Journal on Scientific Computing 19 (6) (1998) 2091–2110.

[16] Jyoti Gupta, Sangeeta Lal, Comparison of some approximate algorithms proposed for traveling salesmen and graph partitioning problems, in: Proceedings of 3rd International Conference on Internet of Things and Connected Technologies, Jaipur (India), March 2018.

[17] Aydın Buluç, Henning Meyerhenke, Ilya Safro, Peter Sanders, Christian Schulz, Recent advances in graph partitioning, in: Algorithm Engineering, Springer, 2016, pp. 117–158.

[18] Marco Chiarandini, Irina Dumitrescu, Thomas Stützle, Stochastic Local Search Algorithms for the Graph Colouring Problem, Handbook of Approximation Algorithms and Metaheuristics, Chapman & Hall, CRC, Boca Raton, FL, USA, 2018.

[19] John J. Hopfield, David W. Tank, Neural computation of decisions in optimization problems, Biological Cybernetics 52 (3) (1985) 141–152.

[20] Xing He, Chuandong Li, Tingwen Huang, Chaojie Li, Junjian Huang, A recurrent neural network for solving bilevel linear programming problem, IEEE Transactions on Neural Networks and Learning Systems 25 (4) (2014) 824–830.

[21] Mohammad Bataineh, Timothy Marler, Neural network for regression problems with reduced training sets, Neural Networks 95 (2017) 1–9.

[22] Mateusz Buda, Atsuto Maki, Maciej A. Mazurowski, A systematic study of the class imbalance problem in convolutional neural networks, Neural Networks 106 (2018) 249–259.

[23] Anh Viet Phan, Minh Le Nguyen, Yen Lam Hoang Nguyen, Dgcnn Lam Thu Bui, A convolutional neural network over large-scale labeled graphs, Neural Networks 108 (2018) 533–543.

[24] Chuangyin Dang, Lei Xu, A globally convergent Lagrange and barrier function iterative algorithm for the traveling salesman problem, Neural Networks 14 (2) (2001) 217–230.

[25] Chuangyin Dang, Yabin Sun, Yuping Wang, Yang Yang, A deterministic annealing algorithm for the minimum concave cost network flow problem, Neural Networks 24 (7) (2011) 699–708.

[26] Chuangyin Dang, Jianqing Liang, Yang Yang, A deterministic annealing algorithm for approximating a solution of the linearly constrained nonconvex quadratic minimization problem, Neural Networks 39 (2013) 1–11.

[27] Chuangyin Dang, Liping He, I.P. Kee Hui, A deterministic annealing algorithm for approximating a solution of the max-bisection problem, Neural Networks 15 (3) (2002) 441–458.

[28] Chuangyin Dang, Wei Ma, Jiye Liang, A deterministic annealing algorithm for approximating a solution of the min-bisection problem, Neural Networks 22 (1) (2009) 58–66.

[29] Babak Jahani, Babak Mohammadi, A comparison between the application of empirical and ANN methods for estimation of daily global solar radiation in Iran, Theoretical and Applied Climatology (2018) 1–13.

[30] Roozbeh Moazenzadeh, Babak Mohammadi, Shahaboddin Shamshirband, Kwokwing Chau, Coupling a firefly algorithm with support vector regression to predict evaporation in northern Iran, Engineering Applications of Computational Fluid Mechanics 12 (1) (2018) 584–597.

[31] Jagat Narain Kapur, Maximum-Entropy Models in Science and Engineering, John Wiley & Sons, 1989.

[32] Irwan Bello, Hieu Pham, Quoc V. Le, Mohammad Norouzi, Samy Bengio, Neural combinatorial optimization, arXiv:1611.09940, 2017.

[33] Richard Durbin, David Willshaw, An analogue approach to the travelling salesman problem using an elastic net method, Nature 326 (6114) (1987) 689.

[34] A. Cochocki, Rolf Unbehauen, Neural Networks for Optimization and Signal Processing, John Wiley & Sons, 1993.

[35] Yanling Wei, Ju H. Park, Hamid Reza Karimi, Yuchu Tian, Hoyoul Jung, Improved stability and stabilization results for stochastic synchronization of continuous-time semi-Markovian jump neural networks with time-varying delay, IEEE Transactions on Neural Networks and Learning Systems 29 (6) (2018) 2488–2501.

[36] John J. Hopfield, Neurons with graded response have collective computational properties like those of two-state neurons, Proceedings of the National Academy of Sciences 81 (10) (1984) 3088–3092.

[37] Michel Minoux, Mathematical Programming: Theory and Algorithms, John Wiley & Sons, 1986.

[38] Zhengtian Wu, Hamid Reza Karimi, Chuangyin Dang, An approximation algorithm for graph partitioning via deterministic annealing neural network, Neural Networks 117 (2019) 191–200.

CHAPTER 8

A logarithmic descent direction algorithm for the quadratic knapsack problem

8.1 Introduction

The famous quadratic knapsack optimization problem introduced in [1] aims to obtain an optimal solution from a quadratic objective function subject to a knapsack constraint and has many applications, such as the least distance [2] and maximum clique problems [3].

The quadratic knapsack problem considered in this study can be expressed as follows:

$$
\begin{aligned}
\min \quad & f(x) = x^T Q x \\
\text{subject to} \quad & \sum_{i=1}^{n} x_i = 1, \\
& 0 \le x_i \le 1, \quad i = 1, 2, \dots, n,
\end{aligned}
\tag{8.1}
$$

where Q is an $n \times n$ positive definite or semipositive definite matrix. From the perspective of computing complexity, problem (8.1) is NP-hard [4,5], and any quadratic minimization problem over a bounded polytope can be transformed into a quadratic knapsack problem, although this transformation is not always practical [6,7].

In recent years, many special knapsack problems have been introduced, and the corresponding algorithms have been developed. For instance, [8] presented a case study of a pseudo-real-life problem and developed a mimetic search approach for an extended quadratic multiple knapsack problem based on a backbone-based crossover operator and a multineighborhood simulated annealing procedure. Meanwhile, [9] extended the single objective quadratic multiple knapsack problem to a biobjective problem and then developed a hybrid two-stage algorithm to approximate the Pareto front of the former. According to the 0–1 knapsack problem, a binary version of the monkey algorithm was developed in [10] to avoid low precision. The grey wolf optimization [11] and modified artificial bee colony approaches [12] have also been used in the literature to address this 0–1 knapsack problem. Meanwhile, [13] introduced the stochastic quadratic

Integer Optimization and Its Computation in Emergency Management
https://doi.org/10.1016/B978-0-32-395203-3.00013-7

multiple knapsack problem and then developed the repair-based approach based on the recoverable robustness technique to solve such a problem. An exact branch-and-price algorithm was developed in [14] to solve the quadratic multiknapsack problem. Meanwhile, [15] introduced the multidemand multidimensional knapsack problem and developed a two-stage search algorithm for solving such a problem. In [16] the dynamic knapsack problem with stochastic item sizes was introduced along with its semiinfinite relaxation. We refer the readers to some other excellent works on special knapsack problems [17–19].

The above research has made excellent results in the area of computation for the knapsack problems. However, all the algorithms have some limitations, and most of them only have ability to solve some special knapsack problems. To the best of the authors' knowledge, there is no algorithm that can effectively solve the quadratic knapsack problem (8.1).

Recently, new algorithms have been developed for this kind of quadratic optimization problems. A globally convergent Lagrange and barrier function iterative algorithm is developed for approximating a solution of the traveling salesman problem in [20]. This algorithm has a good ability to solve the traveling salesman problem with quadratic objective function. Five rather different multistart tabu search strategies for the unconstrained binary quadratic optimization problem are described in [21]. For the NP-hard min-bisection problem, a deterministic annealing algorithm based on a logarithmic-cosine barrier function is developed in [22]. In [23] a set of teaching–learning-based optimization hybrid algorithms is proposed to solve the challenging combinatorial optimization problem, Quadratic Assignment. In [24] an approximate solution of the graph partitioning problem is obtained by using a deterministic annealing neural network algorithm. The algorithm is a continuation method that attempts to obtain a high-quality solution by following a path of minimum points of a barrier problem as the barrier parameter is reduced from a sufficiently large positive number to 0. Another excellent result on computation of quadratic optimization problems can be found in [25,26]. The main contribution of the current study is applying the characters of quadratic optimization to develop a novel logarithmic descent direction algorithm to approximate a solution to the quadratic knapsack problem (8.1). More specifically, the proposed algorithm is developed based on the Karush–Kuhn–Tucker necessary optimality condition and the damped Newton method. Numerical results highlight the effectiveness of the proposed algorithm.

The rest of this chapter is organized as follows. In Section 8.2, we present the proposed logarithmic descent direction algorithm. In Section 8.3, we prove the convergence of the damped Newton method. In Section 8.4, we present numerical experiments to verify the effectiveness of the proposed algorithm. In Section 8.5, we present some concluding remarks.

8.2 Logarithmic descent direction algorithm

Instead of solving (8.1) directly, we consider a scheme for obtaining the solution to (8.1) at a limit as $\beta \uparrow \infty$ of

$$
\begin{aligned}
\min \quad & e(x, \beta) = \beta f(x) - \sum_{j=1}^{n} (\ln x_j + \ln(1 - x_j)) \\
\text{s.t.} \quad & Ex = 1,
\end{aligned}
\tag{8.2}
$$

where $E = [1, 1, \ldots, 1] \in R^n$.

This scheme is taken from Fiacco and McCormick [27] and has been used to solve many optimization problems. When $f(x)$ is convex, this scheme is considered very effective, as reported in the literature. The basic idea of this scheme is as follows. Let β_q, $q = 0, 1, \ldots$, be any sequence such that

$$\beta_0 < \beta_1 < \cdots,$$

and let $\beta_q \to \infty$ as $q \to \infty$. For $q = 0, 1, \ldots$, any available information is used to compute an optimal solution of

$$
\begin{aligned}
\min \quad & e(x, \beta_q) = \beta_q f(x) - \sum_{j=1}^{n} (\ln x_j + \ln(1 - x_j)) \\
\text{s.t.} \quad & Ex = 1.
\end{aligned}
\tag{8.3}
$$

The foundation of this scheme is given in the following lemma (Fiacco and McCormic [27]).

Lemma 8. *Let $x(\beta_q)$ be an optimal solution of (8.3). Then every cluster point of $x(\beta_q)$, $q = 0, 1, \ldots$, is an optimal solution of (8.1).*

For $j = 1, 2, \ldots, n$, let

$$h(x) = \sum_{j=1}^{n} (\ln x_j + \ln(1 - x_j))$$

and

$$L(x, \lambda) = e(x, \beta) + \lambda \left(\sum_{j=1}^{n} x_j - 1 \right).$$

Then

$$\nabla_x L(x, \lambda) = \beta \nabla f(x) + E^T \lambda - \nabla h(x).$$

For any given $\beta \geq 0$, the Karush–Kuhn–Tucker necessary optimality condition states that if x^* is a minimum point of (8.2), then there exists λ^* satisfying

$$\nabla_x L(x^*, \lambda^*) = 0,$$
$$Ex^* = 1, \tag{8.4}$$
$$0 < x^* < 1.$$

When $f(x)$ is convex, (8.4) is a necessary and sufficient condition, and a path-following strategy can be effectively implemented. However, $f(x)$ is nonconvex in this chapter, so that (8.4) is only a necessary condition, and a path-following strategy may fail due to the singularity. When a path-following strategy fails, we can design a procedure for increasing β and use a feasible descent direction method to compute a solution of (8.4) for any given $\beta \geq 0$. We further develop a new feasible descent direction method for computing x^* and λ^* that satisfy (8.4). Letting q_j be the jth row of Q, we obtain

$$\frac{\partial f(x)}{\partial x_j} = 2q_j x.$$

Observe that

$$\frac{\partial L(x, \lambda)}{\partial x_j} = 2\beta q_j x + \lambda - \frac{1}{x_j} + \frac{1}{1 - x_j}.$$

From $\frac{\partial L(x,\lambda)}{\partial x_j} = 0$ we derive

$$x_j = \frac{1}{2} - \frac{1}{2} \frac{\lambda + 2\beta q_j x}{2 + \sqrt{4 + (\lambda + 2\beta q_j x)^2}}, \tag{8.5}$$

$j = 1, 2, \ldots, n$. By substituting (8.5) into $Ex = 1$ we obtain

$$\sum_{j=1}^{n} \left(\frac{1}{2} - \frac{1}{2} \frac{\lambda + 2\beta q_j x}{2 + \sqrt{4 + (\lambda + 2\beta q_j x)^2}} \right) = 1. \tag{8.6}$$

Thus finding x^* and λ^* that satisfy (8.4) is equivalent to finding x^* and λ^* that satisfy (8.5) and (8.6).

Let

$$d_j(x) = \frac{1}{2} - \frac{1}{2}\frac{\lambda + 2\beta q_j x}{2 + \sqrt{4 + (\lambda + 2\beta q_j x)^2}}$$

$j = 1, 2, \ldots, n$, and

$$d(x) = \big(d_1(x), d_2(x), \ldots, d_n(x)\big)^\top.$$

Clearly, $0 < d_j(x) < 1$. The next lemma shows that $d(x) - x$ is a feasible descent direction of $e(x, \beta)$ when $d(x) - x \neq 0$ and $Ed(x) = Ex = 1$.

Lemma 9. *Assume* $0 < x_j < 1$.
- $\frac{\partial L(x,\lambda)}{\partial x_j} > 0$ *if* $d_j(x) - x_j < 0$.
- $\frac{\partial L(x,\lambda)}{\partial x_j} < 0$ *if* $d_j(x) - x_j > 0$.
- $\frac{\partial L(x,\lambda)}{\partial x_j} = 0$ *if* $d_j(x) - x_j = 0$.
- $\nabla_x L(x, \lambda)^\top (d(x) - x) < 0$ *if* $d(x) - x \neq 0$.
- $\nabla e(x, \beta)^\top (d(x) - x) < 0$ *if* $d(x) - x \neq 0$ *and* $Ed(x) = Ex = 1$.

Proof. It suffices to show only that $\frac{\partial L(x,\lambda)}{\partial x_j} > 0$ if $d_j(x) - x_j < 0$. The rest can be obtained similarly.

From $d_j(x) - x_j < 0$ we get

$$\frac{1}{2} - \frac{1}{2}\frac{\lambda + 2\beta q_j x}{2 + \sqrt{4 + (\lambda + 2\beta q_j x)^2}} < x_j. \tag{8.7}$$

Observe that when $\lambda + 2\beta q_j x \neq 0$,

$$\frac{\lambda + 2\beta q_j x}{2 + \sqrt{4 + (\lambda + 2\beta q_j x)^2}} = -\frac{2 - \sqrt{4 + (\lambda + 2\beta q_j x)^2}}{\lambda + 2\beta q_j x} \tag{8.8}$$

and

$$\frac{\lambda + 2\beta q_j x}{2 - \sqrt{4 + (\lambda + 2\beta q_j x)^2}} = -\frac{2 + \sqrt{4 + (\lambda + 2\beta q_j x)^2}}{\lambda + 2\beta q_j x} \tag{8.9}$$

1. Consider the case $\lambda + 2\beta q_j x = 0$. From (8.7) we obtain that $0 < x_j - \frac{1}{2}$. Thus

$$
\begin{aligned}
0 &< \tfrac{2}{x_j(1-x_j)}(x_j - \tfrac{1}{2}) \\
&= -(\tfrac{1}{x_j} - \tfrac{1}{1-x_j}) \\
&= \lambda + 2\beta q_j x - (\tfrac{1}{x_j} - \tfrac{1}{1-x_j}) = \tfrac{\partial L(x,\lambda)}{\partial x_j}.
\end{aligned}
$$

2. Consider the case $\lambda + 2\beta q_j x < 0$. Observe that

$$4 + (\lambda + 2\beta q_j x)^2$$
$$< 4 - 4(\lambda + 2\beta q_j x) + (\lambda + 2\beta q_j x))^2$$
$$= (2 - (\lambda + 2\beta q_j x))^2.$$

Then

$$\sqrt{4 + (\lambda + 2\beta q_j x)^2} < 2 - (\lambda + 2\beta q_j x),$$

and thus

$$\frac{\lambda + 2\beta q_j x}{2 - \sqrt{4 + (\lambda + 2\beta q_j x)^2}} > 1.$$

Therefore

$$x_j - \frac{1}{2} + \frac{1}{2}\frac{(\lambda + 2\beta q_j x)}{2 - \sqrt{4 + ((\lambda + 2\beta q_j x))^2}} > 0.$$

From (8.7) we obtain

$$0 < x_j - \frac{1}{2} + \frac{1}{2}\frac{(\lambda + 2\beta q_j x)}{2 - \sqrt{4 + ((\lambda + 2\beta q_j x))^2}}. \tag{8.10}$$

Multiplying both sides of (8.10) by

$$x_j - \frac{1}{2} + \frac{1}{2}\frac{(\lambda + 2\beta q_j x)}{2 - \sqrt{4 + ((\lambda + 2\beta q_j x))^2}}$$

and using (8.8) and (8.9), we get

$$0 < (x_j - \frac{1}{2} + \frac{1}{2}\frac{(\lambda + 2\beta q_j x)}{2 + \sqrt{4 + ((\lambda + 2\beta q_j x))^2}})$$
$$\times (x_j - \frac{1}{2} + \frac{1}{2}\frac{(\lambda + 2\beta q_j x)}{2 - \sqrt{4 + ((\lambda + 2\beta q_j x))^2}})$$
$$= x_j(x_j - 1) - -\frac{2(x_j - \frac{1}{2})}{\lambda + 2\beta q_j x}$$

Then

$$0 < x_j(x_j - 1) - -\frac{2(x_j - \frac{1}{2})}{\lambda + 2\beta q_j x}. \tag{8.11}$$

Multiplying both sides of (8.11) by $\lambda + 2\beta q_j x$, we get

$$0 > (\lambda + 2\beta q_j x)x_j(x_j - 1) - 2\left(x_j - \frac{1}{2}\right)$$

since $\lambda + 2\beta q_j x < 0$. Thus

$$2\left(x_j - \frac{1}{2}\right) > (\lambda + 2\beta q_j x)x_j(x_j - 1). \qquad (8.12)$$

Dividing both sides of (8.12) by $x_j(x_j - 1)$, we get

$$\lambda + 2\beta q_j x > \frac{2(x_j - \frac{1}{2})}{x_j(x_j - 1)} = \frac{1}{x_j} - \frac{1}{1 - x_j}$$

since $x_j(x_j - 1) < 0$. Therefore

$$0 < \lambda + 2\beta q_j x - \left(\frac{1}{x_j} - \frac{1}{1 - x_j}\right) = \frac{2}{\beta} q_j x \partial L(x, \lambda) \partial x_j.$$

3. Consider the case $\lambda + 2\beta q_j x > 0$. Observe that

$$4 + (\lambda + 2\beta q_j x)^2$$
$$< 4 + 4(\lambda + 2\beta q_j x) + (\lambda + 2\beta q_j x)^2$$
$$= (2 + (\lambda + 2\beta q_j x))^2.$$

Then

$$\sqrt{4 + (\lambda + 2\beta q_j x)^2} < 2 + (\lambda + 2\beta q_j x),$$

and thus

$$\frac{\lambda + 2\beta q_j x}{2 - \sqrt{4 + (\lambda + 2\beta q_j x)^2}} < -1.$$

Therefore

$$x_j - \frac{1}{2} + \frac{1}{2} \frac{\lambda + 2\beta q_j x}{2 - \sqrt{4 + (\lambda + 2\beta q_j x)^2}}$$
$$< x_j - \frac{1}{2} - \frac{1}{2}$$
$$= x_j - 1$$
$$< 0.$$

From (8.7) we obtain

$$0 < x_j - \frac{1}{2} + \frac{1}{2} \frac{\lambda + 2\beta q_j x}{2 + \sqrt{4 + (\lambda + 2\beta q_j x)^2}}. \qquad (8.13)$$

Multiplying both sides of (8.13) by

$$x_j - \frac{1}{2} + \frac{1}{2} \frac{\lambda + 2\beta q_j x}{2 - \sqrt{4 + (\lambda + 2\beta q_j x)^2}}$$

and using (8.8) and (8.9), we get

$$
\begin{aligned}
0 &> (x_j - \tfrac{1}{2} + \tfrac{1}{2} \frac{\lambda + 2\beta q_j x}{2 + \sqrt{4 + (\lambda + 2\beta q_j x)^2}}) \\
&\times (x_j - \tfrac{1}{2} + \tfrac{1}{2} \frac{\lambda + 2\beta q_j x}{2 - \sqrt{4 + (\lambda + 2\beta q_j x)^2}}) \\
&= (x_j - \tfrac{1}{2})^2 - \frac{2(x_j - \tfrac{1}{2})}{\lambda + 2\beta q_j x} - (\tfrac{1}{2})^2 \\
&= x_j(x_j - 1) - \frac{2(x_j - \tfrac{1}{2})}{\lambda + 2\beta q_j x}.
\end{aligned}
$$

Then

$$
0 > x_j(x_j - 1) - \frac{2(x_j - \tfrac{1}{2})}{\lambda + 2\beta q_j x}. \tag{8.14}
$$

Multiplying both sides of (8.14) by $\lambda + 2\beta q_j x$, we get

$$
0 > (\lambda + 2\beta q_j x)x_j(x_j - 1) - 2\left(x_j - \frac{1}{2}\right),
$$

and thus

$$
2\left(x_j - \frac{1}{2}\right) > (\lambda + 2\beta q_j x)x_j(x_j - 1). \tag{8.15}
$$

Dividing both sides of (8.15) by $x_j(x_j - 1)$, we get

$$
\lambda + 2\beta q_j x > \frac{2(x_j - \tfrac{1}{2})}{x_j(x_j - 1)} = \frac{1}{x_j} - \frac{1}{1 - x_j}.
$$

Therefore

$$
0 < \lambda + 2\beta q_j x - \left(\frac{1}{x_j} - \frac{1}{1 - x_j}\right) = \frac{\partial L(x, \lambda)}{\partial x_j},
$$

and the lemma follows.   □

For any given point x, consider

$$
\begin{aligned}
\min \quad & \beta \nabla f(x)^\top \xi \\
\text{subject to} \quad & E\xi = 1, \\
& 0 \le \xi \le 1.
\end{aligned} \tag{8.16}
$$

The analytical center of (8.16) is given by the solution of

$$
\begin{aligned}
\min \quad & \beta \nabla f(x)^\top \xi - \sum_{j=1}^{n}(\ln \xi_j + \ln(1 - \xi_j)) \\
\text{subject to} \quad & E\xi = 1.
\end{aligned} \tag{8.17}
$$

The necessary and sufficient optimality condition states that ξ is the optimal solution of (8.17) if and only if there exists λ that satisfies

$$
\begin{aligned}
&\beta\nabla f(x) + E^{\mathsf{T}}\lambda - (\Xi - S)^{-1}e + (T - \Xi)^{-1}e = 0, \\
&E\xi = 1,
\end{aligned}
\tag{8.18}
$$

where Ξ is a diagonal matrix of ξs, S is a diagonal matrix of 0s, T is a diagonal matrix of 1s, and e is a unit vector. From $\beta\nabla f(x) + E^{\mathsf{T}}\lambda - (\Xi - S)^{-1}e + (T - \Xi)^{-1}e = 0$ we obtain

$$
\xi_j = \frac{1}{2} - \frac{1}{2}\frac{\lambda + 2\beta q_j x}{2 + \sqrt{4 + (\lambda + 2\beta q_j x)^2}}
$$

$j = 1, 2, \ldots, n$. Then, for any given point x, finding a solution of (8.6) is equivalent to finding a solution of (8.18). From the developments of the interior point method [28,29] we know that the solution of (8.18) can be obtained in polynomial time. Thus, we obtain the following:

Lemma 10. *For any given point x, a solution of (8.6) can be obtained in polynomial time.*

From the above discussions we can easily see that for any given $\beta \geq 0$, if x is in a compact set, then λ satisfying (8.18) is in a compact set because E is of full-row rank.

For any given point x, we can also find the solution of (8.6) by the damped Newton method. Let

$$
u_i(\lambda) = \sum_{j=1}^{n}\left(\frac{1}{2} - \frac{1}{2}\frac{\lambda + 2\beta q_j x}{2 + \sqrt{4 + (\lambda + 2\beta q_j x)^2}}\right) - 1,
$$

$i = 1, 2, \ldots, m$, and

$$
u(\lambda) = \left(u_1(\lambda), u_2(\lambda), \ldots, u_m(\lambda)\right)^{\mathsf{T}}.
$$

Let

$$
w(\lambda) = \frac{1}{2}u(\lambda)^{\mathsf{T}}u(\lambda).
$$

By computing the Jacobian matrix $Du(\lambda)$ of $u(\lambda)$ we obtain

$$
Du(\lambda) = -EH(\lambda)E^{\mathsf{T}},
$$

$$H(\lambda) = \begin{pmatrix} h_1(\lambda) & & & \\ & h_2(\lambda) & & \\ & & \ddots & \\ & & & h_n(\lambda) \end{pmatrix},$$

where

$$h_j(\lambda) = \frac{1}{(2 + \sqrt{4 + (\lambda + 2\beta q_j x)^2})\sqrt{4 + (\lambda + 2\beta q_j x)^2}},$$

$j = 1, 2, \ldots, n$.

Observe that $Du(\lambda)$ is negative definite and $\nabla w(\lambda) = Du(\lambda)u(\lambda)$. Let

$$p(\lambda) = -(Du(\lambda))^{-1} u(\lambda).$$

Then, when $u(\lambda) \neq 0$,

$$\begin{aligned} \nabla w(\lambda)^\top p(\lambda) &= -u(\lambda)^\top Du(\lambda)(Du(\lambda))^{-1} u(\lambda) \\ &= -u(\lambda)^\top u(\lambda) < 0, \end{aligned} \tag{8.19}$$

and we compute a solution of (8.6) by the damped Newton method as follows.

Let $\lambda^0 \in R^m$ be an arbitrary point. For $k = 0, 1, \ldots$, use the Cholesky factorization to find the solution $p(\lambda^k)$ of

$$-Du(\lambda^k)p = u(\lambda^k),$$

and let

$$\lambda^{k+1} = \lambda^k + \mu_k p(\lambda^k), \tag{8.20}$$

where $\mu_k \in [0, 1]$ satisfies

$$w(\lambda^k + \mu_k p(\lambda^k)) = \min_{\mu \in [0,1]} w(\lambda^k + \mu p(\lambda^k)).$$

The following lemma is proven in the next section.

Lemma 11. *The damped Newton method converges to the unique solution of (8.6).*

Based on Lemma 9 and the iterative procedure (8.20), we develop a descent direction method for approximating a solution of (8.1) as follows.

Initialization: *Let $\beta = 0$, or let β be a positive number that is sufficiently close to 0, and let $x^0 = E^T/n$. Let λ^0 be an arbitrary point in R^1. Given $x = x^0$, use (8.20) to compute a solution λ^* of (8.6). Meanwhile, let $\lambda^0 = \lambda^*$, and then compute*

$$x_j^1 = \frac{1}{2} - \frac{1}{2} \frac{\lambda^* + 2\beta q_j x_0}{2 + \sqrt{4 + (\lambda^* + 2\beta q_j x_0)^2}},$$

$j = 1, 2, \ldots, n$. Note that $Ex^1 = 1$, let $q = 1$, and go to Step 1.

Step 1:   *Given $x = x^q$, use (8.20) to compute a solution λ^* of (8.6). Let $\lambda^{*,q} = \lambda^*$ and $\lambda^0 = \lambda^*$, and then go to Step 2.*

Step 2:   *Compute*

$$d_j(x^q) = \frac{1}{2} - \frac{1}{2} \frac{\lambda^* + 2\beta q_j x_0}{2 + \sqrt{4 + (\lambda^* + 2\beta q_j x_0)^2}},$$

$j = 1, 2, \ldots, n$. Note that $Ex^q = 1$. If $\|d(x^q) - x^q\|$ is smaller than some given tolerance, then either

- *the method terminates when β is sufficiently large, or*
- *let $\beta = \beta + \Delta\beta$, $x^1 = x^q$, and $q = 1$, and then go to Step 1, where $\Delta\beta$ is a sufficiently small positive number.*

Otherwise, perform the following. Compute

$$x^{q+1} = x^q + \theta_q(d(x^q) - x^q),$$

where $\theta_q \in [0, 1]$, which satisfies

$$e(x^q + \theta_q(d(x^q) - x^q), \beta) = \min_{\theta \in [0,1]} e(x^q + \theta(d(x^q) - x^q), \beta).$$

Let $q = q + 1$, and then go to Step 1.

Theorem 8.   *For any given $\beta \geq 0$, the method converges to a stationary point of (8.2).*

Proof. Let β be any given nonnegative number, and let x^q, $q = 1, 2, \ldots$, be a sequence of points generated by the method. Then

$$e(x^1, \beta) > e(x^2, \beta) > \cdots.$$

We can see from the method that $0 < x^q < 1$, $q = 1, 2, \ldots$. Thus $\lambda^{*,q}$, $q = 1, 2, \ldots$, are bounded, and there is $\delta > 0$ such that $\delta e \leq x^q \leq 1 - \delta e$, $q = 1, 2, \ldots$.

Let x^{q_k}, $k = 1, 2, \ldots$, be a convergent subsequence of x^q, $q = 1, 2, \ldots$, and let x^* be the limit of x^{q_k}, $k = 1, 2, \ldots$. Observe that $\delta e \leq x^* \leq 1 - \delta e$. Note that when $0 < x < 1$ and $Ad(x) = Ax = b$, $d(x) - x = 0$ if and only if $\nabla e(x, \beta) = 0$. Suppose that $\nabla e(x^*, \beta) \neq 0$. Therefore $d(x^*) - x^* \neq 0$, and we have

$$\nabla e(x^*, \beta)^\top (d(x^*) - x^*) < 0.$$

Let $x^{**} = x^* + \theta^*(d(x^*) - x^*)$, where θ^* is a number in $[0, 1]$ that satisfies

$$e(x^* + \theta^*(d(x^*) - x^*), \beta) = \min_{\theta \in [0,1]} e(x^* + \theta(d(x^*) - x^*), \beta).$$

Then $e(x^{**}, \beta) < e(x^*, \beta)$, and let $\varepsilon = e(x^*, \beta) - e(x^{**}, \beta) > 0$. Given that $e(x, \beta)$ is continuous, there exists $\delta > 0$ such that

$$|e(x, \beta) - e(x^{**}, \beta)| < \frac{\varepsilon}{2}$$

when $\|x - x^{**}\| < \delta$. Choose a sufficiently large q_k such that

$$\|x^{q_k} - x^*\| < \frac{\delta}{2}$$

and

$$\|d(x^{q_k}) - x^{q_k} - (d(x^*) - x^*)\| < \frac{\delta}{2\theta^*},$$

given that $x^{q_k} \to x^*$ as $k \to \infty$ and $d(x) - x$ is continuous. Therefore

$$\begin{aligned}
&\|x^{q_k} + \theta^*(d(x^{q_k}) - x^{q_k}) - x^{**}\| \\
&= \|x^{q_k} + \theta^*(d(x^{q_k}) - x^{q_k}) - (x^* + \theta^*(d(x^*) - x^*))\| \\
&\leq \|x^{q_k} - x^*\| + \theta^*\|d(x^{q_k}) - x^{q_k} - (d(x^*) - x^*)\| \\
&< \tfrac{\delta}{2} + \theta^*\tfrac{\delta}{2\theta^*} \\
&= \delta,
\end{aligned}$$

and

$$\begin{aligned}
e(x^{q_k+1}, \beta) &= \min_{\theta \in [0,1]} e(x^{q_k} + \theta(d(x^{q_k}) - x^{q_k}), \beta) \\
&\leq e(x^{q_k} + \theta^*(d(x^{q_k}) - x^{q_k}), \beta) \\
&\leq e(x^{**}, \beta) + \tfrac{\varepsilon}{2} \\
&= e(x^{**}, \beta) + \tfrac{e(x^*, \beta) - e(x^{**}, \beta)}{2} \\
&< e(x^*, \beta) \\
&< e(x^{q_k+1}, \beta) \\
&\leq e(x^{q_k+1}, \beta),
\end{aligned}$$

where the last inequality comes from the facts that of $q_k + 1 \leq q_{k+1}$ and $e(x^q, \beta)$ decreases monotonically. Hence we have a contradiction. The proof is complete. □

It is well known that the predictor–corrector method is efficient for solving linearly constrained convex minimization problems. When β is small, we can see that $e(x, \beta)$ is convex. Hence the predictor–corrector method may be used whenever it is possible. A detailed discussion about the predictor–corrector method can be found in [30].

Consider

$$\beta \nabla f(x) - (X - S)^{-1}e + (T - X)^{-1}e + E^\top \lambda = 0,$$
$$Ex = 1.$$
(8.21)

By differentiating (8.21) with respect to β we obtain

$$\begin{pmatrix} \beta Q + (X-S)^{-2} + (T-X)^{-2} & E^\top \\ E & 0 \end{pmatrix} \begin{pmatrix} \frac{dx}{d\beta} \\ \frac{d\lambda}{d\beta} \end{pmatrix}$$
$$= - \begin{pmatrix} \nabla f(x) \\ 0 \end{pmatrix}.$$

Let (x^0, λ^0) be the solution of

$$-(X - S)^{-1}e + (T - X)^{-1}e + E^\top \lambda = 0,$$
$$Ex = 1.$$

Then the matrix

$$\begin{pmatrix} (X_0 - S)^{-2} + (T - X_0)^{-2} & E^\top \\ E & 0 \end{pmatrix}$$

is invertible. The implicit function theorem states that the equation

$$\begin{pmatrix} \beta Q + (X-S)^{-2} + (T-X)^{-2} & A^\top \\ A & 0 \end{pmatrix} \begin{pmatrix} \frac{dx}{d\beta} \\ \frac{d\lambda}{d\beta} \end{pmatrix}$$
$$= - \begin{pmatrix} \nabla f(x) \\ 0 \end{pmatrix},$$
(8.22)

$$x(0) = x^0,$$
$$\lambda(0) = \lambda^0$$

has a solution on some interval $[0, \beta^*)$ with $\beta^* > 0$. Let $x(\beta)$, $\beta \in [0, \beta^*)$, be the solution. Then $\beta Q + (X(\beta) - S)^{-2} + (T - X(\beta))^{-2}$ is positive definite on $[0, \beta^*)$. Thus, for any $\beta \in (0, \beta^*)$, if $x(\beta) \neq x^0$, then

$$\nabla f\left(x(\beta)\right)^\top \frac{dx}{d\beta} < 0.$$

Lemma 12. *Assume that $x(\beta) \neq x^0$, $\beta \in (0, \beta^*)$. For $\beta \in [0, \beta^*)$, $f(x(\beta))$ decreases monotonically as β increases.*

If β^* reaches infinity, then the predictor–corrector method is an efficient procedure for approximating a solution of (8.1). However, β^* may also be finite. In this case, we develop a hybrid method that applies the descent direction method whenever the predictor–corrector method is inapplicable.

8.3 Convergence of the damped Newton method

In this section, we illustrate that the iterative procedure (8.20) converges to the unique solution λ^* of (8.6). We first prove that for any i, no subsequence of λ_i^k, $k = 0, 1, \ldots$, approaches ∞ or $-\infty$. As a consequence, we find that as $k \to \infty$, $u(\lambda^k)$ and $p(\lambda^k)$ both approach zero. We then deduce that as $k \to \infty$, λ^k converges to λ^*.

Note that P has an interior point. Let $\bar{x}$ be an interior point of P, and let $E\bar{x} = 1$ and $0 < \bar{x} < 1$. Thus, for any $\lambda \in R^1$, we have

$$\lambda = (E\bar{x})^\top \lambda = \bar{x}^\top E^\top \lambda = \sum_{j=1}^{n} \bar{x}_j \lambda. \tag{8.23}$$

If

$$\xi(\lambda) = \sum_{j=1}^{n} \int_0^\lambda \left(\frac{1}{2} - \frac{1}{2} \frac{r + 2\beta q_j x}{2 + \sqrt{4 + (r + 2\beta q_j x)^2}} \right) dr - \lambda, \tag{8.24}$$

then we have

$$\nabla \xi(\lambda)$$
$$= \sum_{j=1}^{n} a_j \left(\frac{1}{2} - \frac{1}{2} \frac{a_j^\top \lambda + 2\beta q_j x}{2 + \sqrt{4 + (\lambda + 2\beta q_j x)^2}} \right) - 1$$
$$= u(\lambda).$$

Given that $-Du(\lambda)$ is positive definite, we have

$$\nabla \xi(\lambda)^\top p(\lambda) = -u(\lambda) \left(Du(\lambda) \right)^{-1} u(\lambda) > 0.$$

Therefore $\xi(\lambda^k)$ is nondecreasing because $\lambda^{k+1} = \lambda^k + \mu_k p(\lambda^k)$. From the mean-value integration theorem we obtain

$$\begin{aligned}\sum_{j=1}^{n} \int_0^\lambda (\tfrac{1}{2} - \tfrac{1}{2}\frac{r+2\beta q_j x}{2+\sqrt{4+(r+2\beta q_j x)^2}})dr \\ = \sum_{j=1}^{n}(\tfrac{1}{2} - \tfrac{1}{2}\frac{r+2\beta q_j x}{2+\sqrt{4+(r+2\beta q_j x)^2}}),\end{aligned} \tag{8.25}$$

where $r_j(\lambda)$ is a number between 0 and λ that satisfies $r_j(\lambda) \to \infty$ as $\lambda \to \infty$ and $r_j(\lambda) \to -\infty$ as $\lambda \to -\infty$. By substituting (8.23) and (8.25) into (8.24) we obtain

$$\xi(\lambda) = \sum_{j=1}^{n}\left(\frac{1}{2} - \frac{1}{2}\frac{r+2\beta q_j x}{2+\sqrt{4+(r+2\beta q_j x)^2}} - \bar{x}_j\right)\lambda. \tag{8.26}$$

Given that $\xi(\lambda^k)$ is nondecreasing, we find from (8.26) that for any j, no subsequence of λ^k, $k = 0, 1, \ldots$, approaches ∞ or $-\infty$. Therefore there are two finite constants L and U such that for any k, we have

$$L \le \lambda^k \le U,$$

$k = 1, 2, \ldots, n$.

Suppose that there is l for which a subsequence of λ_l^k, $k = 0, 1, \ldots$, approaches ∞ or $-\infty$. Let $\lambda_l^{k_v}$, $v = 0, 1, \ldots$, be such a subsequence. Then, for any v, we have

$$L \le \lambda^{k_v} \le U, \tag{8.27}$$

$k = 1, 2, \ldots, n$. By dividing both sides of (8.27) by $\lambda_l^{k_v}$ we obtain

$$\frac{L}{\lambda_l^{k_v}} \le \sum_{i\ne l}\frac{\lambda_i^{k_v}}{\lambda_l^{k_v}} + 1 \le \frac{U}{\lambda_l^{k_v}}, \tag{8.28}$$

$k = 1, 2, \ldots, n$. By letting v approach infinity, we find from (8.28) that there is a nonzero vector z that satisfies $E^{\mathsf{T}} z = 0$, thereby contradicting the finding that A is of full-row rank. Therefore, for any i, no subsequence of λ_i^k, $k = 0, 1, \ldots$, approaches ∞ or $-\infty$.

We observe that $\nabla w(\lambda^k)^{\mathsf{T}} p(\lambda^k) < 0$ if $u(\lambda^k) \ne 0$. Given the boundedness of λ^k, $k = 0, 1, \ldots$, we find that $u(\lambda^k)$ approaches zero as $k \to \infty$. Therefore λ^k converges to the unique solution of (8.6). The convergence of the iterative procedure (8.20) is proven.

8.4 Numerical results

In this section, we present the experimental results for some quadratic knapsack problems to validate the effectiveness of the proposed approach. The experimental problems are generated randomly, and the algorithm is programmed in MATLAB®. As a comparison, we also examine the barrier algorithm (BA) [17], which has got the best application in industry. The BA algorithm has been embedded in the CPLEX software for the nonconvex quadratic programming problems.

All problems take the following form:

$$\begin{aligned}
\min \quad & f(x) = x^T Q x \\
\text{subject to} \quad & \sum_{i=1}^{n} x_i = 1, \\
& 0 \le x_i \le 1, i = 1, 2, \ldots, n,
\end{aligned} \qquad (8.29)$$

where Q is a randomly generated $n \times n$ a positive definite or semipositive symmetric matrix. In the implementation of the algorithm, β is initially equal to 0.001 and increases by a factor of 1.05 when $\|d(x^q) - x^q\| < 0.001$.

Example 9. We generate this example randomly, with the following data: $n = 5$, and

$$Q = \begin{pmatrix}
55 & 11 & 61 & 48 & 72 \\
11 & 59 & 80 & 99 & 40 \\
61 & 80 & 90 & 29 & 46 \\
48 & 99 & 29 & 12 & 24 \\
72 & 40 & 46 & 24 & 18
\end{pmatrix}.$$

We apply the proposed algorithm and the BA algorithm using the above data. The numerical result shows that the proposed algorithm converges to the optimal solution $x^* = (0, 0, 0, 1, 0)^T$ after 267 iterations with an object value 12.

In this example, the BA algorithm can also obtain the same solution with the same objective. The computational complexity analysis is an important procedure for validating the effectiveness of an algorithm. However, for certain algorithms such as the proposed algorithm and the annealing and genetic algorithms, we can only get the number of steps in every iteration, and it is very hard to obtain the exact number of iterations, as well as the total number of steps. Therefore, analyzing the computational complexity of our method is very hard. To deal with this problem, Figs. 8.1 and 8.2 describe the convergent trajectories of the variables and the state trajectories

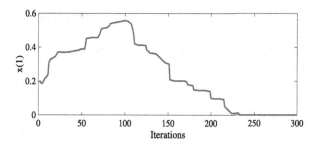

Figure 8.1 Convergent trajectories of the first state variables for Example 9.

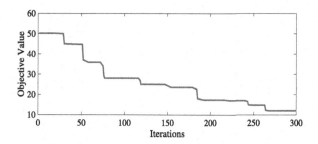

Figure 8.2 Convergent trajectories of the objective function for Example 9.

of the objective function, respectively. The figures show that the algorithm can converge to the optimal solution in Example 9.

Example 10. We randomly generate 30 examples with the following data: $n = 20$, and Q is a randomly generated 20×20 positive definite or semipositive symmetric matrix. The proposed algorithm and the BA algorithm are successively applied to these 30 examples. The objective functions solved by the proposed method and the BA method are given in Table 8.1.

To compare the proposed algorithm with the BA algorithm properly at the same level, we define

$$comp = \frac{\text{BA method} - \text{the proposed method}}{\text{the proposed method}} \times 100\%, \qquad (8.30)$$

where *comp* stands for the performance of the proposed method compared with the BA method. A positive *comp* means that the proposed method is better than the BA method. The greater the *comp*, the better the performance of the proposed method.

Table 8.1 Object function values of 30 examples using the proposed method and the BA method.

Index	The proposed method	BA method	Index	The proposed method	BA method
1	3323	3577	16	7346	7980
2	2972	3410	17	6244	8010
3	3571	4288	18	6234	7902
4	2120	3453	19	2344	3490
5	1989	1989	20	4434	5625
6	3753	4888	21	3520	4521
7	4509	5398	22	4523	4523
8	4289	6081	23	5632	6733
9	2045	3087	24	3342	5320
10	2794	3098	25	3860	3985
11	5823	7010	26	4522	5796
12	2099	2980	27	2455	3482
13	3409	4109	28	3254	4822
14	1929	2269	29	3459	4531
15	3453	3922	30	4255	4870

Let *comp* be on the vertical axis, and let the 30 test examples be distributed on the horizontal axis. The comparisons of the proposed method with the BA method are illustrated in Fig. 8.3.

As we see in Fig. 8.3, the proposed method has a better objective function than the BA method for 28 test examples and the same values for 2 test examples. Overall, according to the value of the vertical axis, the proposed method is 30% better than the BA method on average.

From the above examples we see that the proposed logarithmic descent direction algorithm is effective for the quadratic knapsack problem. However, the performance of our deterministic annealing neural network algorithm partly depends on the selected parameters, such as β, increasing factor, and so on. The optimal values of these parameters cannot be calculated theoretically and can only be obtained from experience and experiment. The experiment will cost some time before computation. Besides, for the proposed algorithm, we can only obtain the computational complexity in every iteration, but the total number of iterations could not be obtained. Therefore it is hard to make the computational complexity analysis of the proposed algorithm.

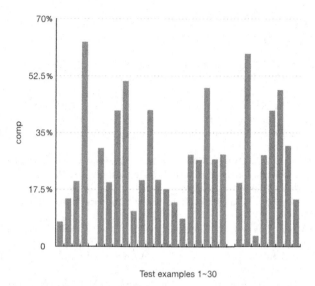

Test examples 1~30

Figure 8.3 Comparison of the object function values of the proposed method and the BA method.

8.5 Conclusions

The computation of the knapsack problem is a hot topic in both theory and practice. In this study, we focus on the quadratic knapsack problem, which is NP-hard. In this chapter, we develop a logarithmic descent direction algorithm to approximate a solution for the quadratic knapsack problem. To construct this algorithm, we apply the Karush–Kuhn–Tucker necessary optimality condition to transform the formulation of the problem. However, the solution still cannot be obtained by the reformulation because the quadratic knapsack problem may be a nonconvex problem. Based on this reformulation, to solve the reformulation, we develop a damped Newton method and an iterative procedure, which we prove to be globally convergent. The simulation results confirm the effectiveness of the proposed algorithm for solving the quadratic knapsack problem. In the future work, the applicability of the algorithm will be paid more attention, and more computing technologies will be considered to improve the algorithm. The details of the algorithm can be found in [31].

References

[1] Giorgio Gallo, Peter L. Hammer, Bruno Simeone, Quadratic knapsack problems, in: Combinatorial Optimization, Springer, 1980, pp. 132–149.

[2] Philip Wolfe, Algorithm for a least-distance programming problem, in: Pivoting and Extension, Springer, 1974, pp. 190–205.

[3] Theodore S. Motzkin, Ernst G. Straus, Maxima for graphs and a new proof of a theorem of Turán, Canadian Journal of Mathematics 17 (1965) 533–540.

[4] Michael R. Gary, David S. Johnson, Computers and Intractability: A Guide to the Theory of NP-Completeness, W.H. Freeman and Company, New York, 1979.

[5] Panos M. Pardalos, Georg Schnitger, Checking local optimality in constrained quadratic programming is NP-hard, Operations Research Letters 7 (1) (1988) 33–35.

[6] Panos M. Pardalos, Judah Ben Rosen, Constrained Global Optimization: Algorithms and Applications, vol. 268, Springer, 1987.

[7] Panos M. Pardalos, Yinyu Ye, Chi-Geun Han, Algorithms for the solution of quadratic knapsack problems, Linear Algebra and Its Applications 152 (1991) 69–91.

[8] Yuning Chen, Jin-Kao Hao, Memetic search for the generalized quadratic multiple knapsack problem, IEEE Transactions on Evolutionary Computation 20 (6) (2016) 908–923.

[9] Yuning Chen, Jinkao Hao, The bi-objective quadratic multiple knapsack problem: model and heuristics, Knowledge-Based Systems 97 (2016) 89–100.

[10] Yongquan Zhou, Xin Chen, Guo Zhou, An improved monkey algorithm for a 0–1 knapsack problem, Applied Soft Computing 38 (2016) 817–830.

[11] Eman Yassien, Raja Masadeh, Abdullah Alzaqebah, Ameen Shaheen, Grey wolf optimization applied to the 0/1 knapsack problem, International Journal of Computer Applications 169 (5) (2017) 11–15.

[12] Jie Cao, Baoqun Yin, Xiaonong Lu, Yu Kang, Xin Chen, A modified artificial bee colony approach for the 0–1 knapsack problem, Applied Intelligence (2018) 1–14.

[13] Bingyu Song, Yanling Li, Yuning Chen, Feng Yao, Yingwu Chen, A repair-based approach for stochastic quadratic multiple knapsack problem, Knowledge-Based Systems 145 (2018) 145–155.

[14] David Bergman, An exact algorithm for the quadratic multiknapsack problem with an application to event seating, INFORMS Journal on Computing 31 (3) (2019) 477–492.

[15] Xiangjing Lai, Jin-Kao Hao, Dong Yue, Two-stage solution-based tabu search for the multidemand multidimensional knapsack problem, European Journal of Operational Research 274 (1) (2019) 35–48.

[16] Daniel Blado, Alejandro Toriello, Relaxation analysis for the dynamic knapsack problem with stochastic item sizes, SIAM Journal on Optimization 29 (1) (2019) 1–30.

[17] John M. Mellor-Crummey, Michael L. Scott, Algorithms for scalable synchronization on shared-memory multiprocessors, ACM Transactions on Computer Systems (TOCS) 9 (1) (1991) 21–65.

[18] Pierre Bonami, Andrea Lodi, Jonas Schweiger, Andrea Tramontani, Solving quadratic programming by cutting planes, SIAM Journal on Optimization 29 (2) (2019) 1076–1105.

[19] Mohamed Abdel-Basset, Doaa El-Shahat, Hossam Faris, Seyedali Mirjalili, A binary multi-verse optimizer for 0–1 multidimensional knapsack problems with application in interactive multimedia systems, Computers & Industrial Engineering 132 (2019) 187–206.

[20] Chuangyin Dang, Lei Xu, A globally convergent Lagrange and barrier function iterative algorithm for the traveling salesman problem, Neural Networks 14 (2) (2001) 217–230.

[21] Gintaras Palubeckis, Multistart tabu search strategies for the unconstrained binary quadratic optimization problem, Annals of Operations Research 131 (1–4) (2004) 259–282.

[22] Chuangyin Dang, Wei Ma, Jiye Liang, A deterministic annealing algorithm for approximating a solution of the min-bisection problem, Neural Networks 22 (1) (2009) 58–66.

[23] Tansel Dokeroglu, Hybrid teaching–learning-based optimization algorithms for the quadratic assignment problem, Computers & Industrial Engineering 85 (2015) 86–101.

[24] Zhengtian Wu, Hamid Reza Karimi, Chuangyin Dang, An approximation algorithm for graph partitioning via deterministic annealing neural network, Neural Networks 117 (2019) 191–200.

[25] Yurii Nesterov, Lectures on Convex Optimization, vol. 137, Springer, 2018.

[26] Jie Han, Chunhua Yang, Xiaojun Zhou, Weihua Gui, A two-stage state transition algorithm for constrained engineering optimization problems, International Journal of Control, Automation, and Systems 16 (2) (2018) 522–534.

[27] Anthony V. Fiacco, Garth P. McCormick, Nonlinear Programming: Sequential Unconstrained Minimization Techniques, vol. 4, SIAM, 1990.

[28] Yinyu Ye, On homogeneous and self-dual algorithms for LCP, Mathematical Programming 76 (1) (1997) 211–221.

[29] Paul Tseng, Yinyu Ye, On some interior-point algorithms for nonconvex quadratic optimization, Mathematical Programming 93 (2) (2002) 217–225.

[30] Eugene L. Allgower, Kurt Georg, et al., Computational Solution of Nonlinear Systems of Equations, vol. 26, American Mathematical Soc., 1990.

[31] Zhengtian Wu, Baoping Jiang, Hamid Reza Karimi, A logarithmic descent direction algorithm for the quadratic knapsack problem, Applied Mathematics and Computation 369 (2020) 124854.

Index

Printed in the United States
by Baker & Taylor Publisher Services